STUDY GUIDE

to accompany

Chemistry and the Living Organism

FIFTH EDITION

Molly M. Bloomfield

John Wiley & Sons, Inc.

New York • Chichester • Brisbane • Toronto • Singapore

ISBN 0-471-51807-7

Printed in the United States of America
Printed and bound by the Hamilton Printing Company.
10 9 8 7 6 5 4 3 2

Preface

The organization of this study guide to the fifth edition of
CHEMISTRY AND THE LIVING ORGANISM carefully follows the format of
the textbook. Each chapter begins with a summary of the
important concepts covered, and then has a section-by-section
discussion of these topics. For each section there is a list of
important terms that appear in that section. Each chapter
contains many self-test questions, whose answers are found at the
end of the chapter, to help the student measure her or his
comprehension of the topics covered. Appendix I contains the
mathematical solutions to the in-chapter exercises in the text.
New in this edition of the study guide is Appendix II, which
contains the answers to the end-of-chapter review problems in the
text.

Contents

To The Student

This study guide accompanies CHEMISTRY AND THE LIVING ORGANISM,
Fifth Edition. I wrote it to help you better understand the
principles presented in the text. The study guide will assist
you in picking out the important concepts in each chapter, and
will aid your review when studying for exams.

The study guide begins with a mathematics review that will help
those of you whose math might be a bit rusty. This review
reinforces the material found in Appendices 1 and 2 and in
Sections 1.7 through 1.10 of your text.

The study guide closely follows the topics in the text. You can
use both books together to increase the effectiveness of your
study time. Each chapter in the study guide begins with a
paragraph summarizing the topics covered in that chapter. If you
read this summary before reading the chapter in the text, it will
help you to identify the important material to be covered. The
chapters in the study guide are divided into sections matching
the sections in the text. After you have read through a complete
chapter in the text for the first time, begin to read it again
section-by-section. As you do, also read the section summaries
in the study guide. Write out a definition in your own words for
each of the important terms listed in the guide. To help you
with these definitions, there is a Glossary of all important
terms at the end of your text. Check your understanding of the
material by answering the Self-Test questions in the study guide.
The answers to all of these questions are found at the end of
each chapter in the guide. In addition, I have included at the
end of the study guide the answers to the Review Questions in the
text, as well as worked-out solutions to the in-chapter
Exercises. Your class notes and this study guide should provide
excellent review materials for quizzes and exams.

I hope this study guide will make your study of chemistry more
understandable and enjoyable.

Molly M. Bloomfield

Introduction   MATHEMATICS REVIEW

In this chemistry course, you will often be asked to solve numerical problems. This math review describes the basic mathematical principles you must know to solve such problems. You will find it helpful to study this section carefully before beginning the exercises in the Study Guide. You might want to review it again before beginning Chapters 5, 6, 11, and 12 in the text. This math review is meant to go along with the material covered in Sections 1.7-1.10 and Appendices 1 and 2 in the text.

Numbers in Chemistry

You will find that the numbers used in chemical calculations are expressed not only as whole numbers (1, 8, 295), but also as fractions (8/11. 194/278), decimals (8.45, 413.8), or numbers in exponential form (8×10^{-2}, 2.37×10^{12}). Most numbers will be positive, but a few will be negative numbers (-8, -23.67) with values less than zero. Numbers used in chemical calculations are usually measurements. Therefore, they should have some unit of measurement attached to them so that we know what is being measured (8.45 centimeters, 478 grams).

Fractions and Decimals

Fractions can be converted into decimal numbers be performing the division indicted by the fraction. That is, 3/4 actually means "3 divided by 4", and if we divide 4 into 3 we get the answer 0.75. Thus, the fraction 3/4 is the same as the decimal number 0.75.

Numbers Raised to Powers

When a number is multiplied by itself, that number is said to be squared. This is represented by writing a superscript of 2 above and to the right of the number:

$$6 \times 6 = 6^2 = 36$$

To cube a number means to multiply that number by itself 3 times:

$$(1.3)^3 = (1.3) \times (1.3) \times (1.3) = 2.197$$

The power of a number tells us how many times the number is to be multiplied times itself: 1.3 to the power 3 (1.3 to the third) equals 2.197.

Numbers in Exponential Form

Appendix 1 in the text explains how to write numbers in
exponential form. Read this Appendix carefully, and then try the
following problems.

Self-Test _____

For questions 1 to 10, write the number in exponential form.

1. 1560 1.560×10^3
2. 7,432,000 7.432×10^6
3. 0.0046 4.6×10^{-3}
4. 145 1.45×10^2
5. 0.025 2.5×10^{-2}

6. 0.0000007 7.0×10^{-7}
7. 456,000 4.56×10^5
8. 0.000041 4.1×10^{-5}
9. 25,000 2.5×10^4
10. 0.15 1.5×10^{-1}

If you had difficulty with this self-test, study the following
discussion and then try the self-test again.

A number expressed in exponential form is written as a number
between 1 and 10 (called the coefficient), multiplied by the
number 10 raised to some power. Consider, for example, the
number 2165. You know that this number has a value that is just
a little greater than 2 times the number 1000. But the number
1000 is the number 10 to the third power:

$$1000 = 10 \times 10 \times 10 = 10^3$$

Thus, 2165 is just a little more than 2 times 10 to the third;
specifically, $2165 = 2.165 \times 10^3$. Notice that to change 2165 to
2.165 we had to move the decimal point 3 positions to the left --
but each one of these decimal place moves corresponds to
changing the number by a power of ten. For our final answer to
have the same numerical value as 2165, for each place that we
move the decimal point to the left, we must multiply by ten.
Thus, "three moves to the left" requires that we multiply our
final number by $10 \times 10 \times 10$ or 10^3.

From this example we can state simple rules for changing numbers
into their exponential form:

1.   First, count the number of places that the decimal point
     must be moved to create a coefficient between 1 and 10.

     | 246,000 | 2.46000 | 2.46 | 5 places to the left |
     | 0.00485 | 4.85 | 4.85 | 3 places to the right |

2. The power of 10 to use is equal to the number of places that the decimal point was moved. However, if the original number was less than 1.0 (and it was necessary to move the decimal point to the right), a minus sign is written in front of the power of 10.

$$246,000 = 2.46 \times 10^5$$
$$0.00485 = 4.85 \times 10^{-3}$$

Now try the self-test questions 1 to 10 again, following these rules.

Mathematical Operations

Addition and Subtraction

You will rarely need to add or subtract fractions from one another in chemical calculations, but you will be required to add or subtract numbers expressed as decimals or in exponential form. No matter in what form the numbers appear, however, the most important thing to check is that you are adding or subtracting numbers that represent exactly the same type of units. For example, it makes no sense to add 6 inches to 3 feet and to claim that you end up with 6 + 3 = 9 of something. Either you must express both measurements in inches (6 inches + 36 inches = 42 inches) or in feet (0.5 feet + 3 feet = 3.5 feet). Section 1.10 describes how to use unit factors to convert numbers so that they all represent the same kind of units.

Decimal Numbers. To add or subtract numbers expressed as decimals, the only rule is to line up all the decimal points one under the other before you start adding or subtracting. The decimal point in the answer will appear under all the other decimal points.

Example

1. Add the following numbers: 3.4, 0.026, 274.35, 10.458, and 0.14.

   Lining up all the decimal points one under the other, we have

$$
\begin{array}{r}
3.4 \\
0.026 \\
274.35 \\
10.458 \\
\underline{0.14} \\
288.374
\end{array}
$$

2.   Subtract 10.476 from 283.9

Lining up the decimal points, we have

$$283.9$$
$$\underline{-10.476}$$

Because adding zeros to the numbers on the right of the decimal point doesn't change the number, we can rewrite the problem in a form that is easier to work with.

$$283.900$$
$$\underline{-10.476}$$
$$273.424$$

Numbers in Exponential form.   To add or subtract numbers in exponential form, all numbers must first be written to the same power of 10 (that is, all the exponents of the number 10 must be the same).   If you are unsure of how to perform that operation, see the section in this math review that discusses multiplication of numbers in exponential form.   Once all the numbers are written to the same power of 10, the coefficients of the numbers are added or subtracted according to the procedure we just discussed for decimal numbers.

Example _____

1.   Add 1.043×10^3 to 5.62×10^3

Because both numbers have the same power of 10, we can just add the two numbers together (being sure to line up the decimal points).

$$1.043 \times 10^3$$
$$\underline{5.62 \ \times 10^3}$$
$$6.663 \times 10^3$$

2.   Add 6.29×10^4, 1.79×10^6, and 25.77×10^4.

We must first rewrite the numbers so that they all have the same power of ten.   One way to do this is to rewrite

$$1.79 \times 10^6 = 1.79 \times 10^2 \times 10^4 = 179 \times 10^4$$

Then, lining up the decimal points and adding zeros to the right of the decimal point, we have

$$6.29 \times 10^4$$
$$179.00 \times 10^4$$
$$\underline{25.77 \times 10^4}$$
$$211.06 \times 10^4$$

4

3. Subtract 7.47×10^{-5} from 2.16×10^{-4}.

First, rewrite $2.16 \times 10^{-4} = 2.16 \times 10 \times 10^{-5} = 21.6 \times 10^{-5}$.
Then,

$$
\begin{array}{r}
21.60 \times 10^{-5} \\
- \ 7.47 \times 10^{-5} \\
\hline
14.13 \times 10^{-5}
\end{array}
$$

Self-Test

For questions 11 to 20, perform the indicated calculations.

11. $25.4 + 0.35 + 5.3$ *31.05*
12. $0.1 + 0.004 + 0.0005$ *0.1045*
13. $320.1 \times 10^{6} + 0.02 \times 10^{9}$ *3.401 ^{09}*
14. $46.03 + 140 + 0.2 + 0.0015$
15. $45 \times 10^{-5} + 3.6 \times 10^{-3}$ *0.0405 3.210 ^{09} 186.2315*

16. $145 - 0.136$ *144.864*
17. $6.485 - 0.01$ *6.475*
18. $48.0 \times 10^{6} - 360 \times 10^{5}$ *1.208*
19. $1.2 \times 10^{-8} - 60 \times 10^{-6}$ *-0.0005998*
20. $6 \times 10^{-8} - 0.45 \times 10^{-7}$ *.0000001*

Multiplication

The multiplication of a number "a" times another number "b" can be expressed in several ways:

$$
a \times b, \quad a(b), \quad (a)(b), \quad \text{or} \quad \frac{a}{xb}
$$

Fractions. When several fractions are multiplied together, the result will be another fraction whose top part (or numerator) is the product of all the other numerators multiplied together, and whose bottom part (or denominator) is the product of all the other denominators multiplied together. If the various numbers that appear in the fractions also have units of measure attached to them, the units of measure are also multiplied together to form a "fraction" made up of various units of measure. THese units will cancel each other out if they appear in both the numerator and the denominator.

Example

1. Multiply $\frac{4}{5}$ times $\frac{3}{7}$.

$$
\frac{4}{5} \times \frac{3}{7} = \frac{4 \times 3}{5 \times 7} = \frac{12}{35}
$$

2. Multiply together $\frac{1}{5}$, $\frac{2}{3}$, $\frac{7}{12}$ and $\frac{5}{8}$.

5

$$\frac{1}{5} \times \frac{2}{3} \times \frac{7}{12} \times \frac{5}{8} = \frac{1 \times 2 \times 7 \times 5}{5 \times 3 \times 12 \times 8} = \frac{70}{1440} = \frac{7}{144}$$

3. $\quad 15 \text{ yd} \times \dfrac{3 \text{ ft}}{1 \text{ yd}} \times \dfrac{12 \text{ in}}{1 \text{ ft}} = \dfrac{15 \times 3 \times 12}{1 \times 1} \quad \dfrac{\text{yd} \times \text{ft} \times \text{in}}{\text{yd} \times \text{ft}} = 540 \text{ in}$

4. $\quad \dfrac{6.3 \text{ grams}}{3 \text{ moles}} \times \dfrac{2 \text{ moles}}{1 \text{ liter}} \times \dfrac{1 \text{ liter}}{1000 \text{ milliliters}}$

$\quad = \left(\dfrac{6.3 \times 2 \times 1}{3 \times 1 \times 1000} \right) \left(\dfrac{\text{grams} \times \text{moles} \times \text{liter}}{\text{moles} \times \text{liter} \times \text{milliliters}} \right)$

$\quad = \dfrac{12.6 \text{ grams}}{3000 \text{ milliliters}} = 0.0042 \dfrac{\text{grams}}{\text{milliliters}}$

Decimal Numbers. When several decimal numbers are multiplied together, the result is another decimal number. The number of decimal places in the answer (that is, the number of figures appearing to the right of the decimal point) equals the sum of the number of decimal places of all the numbers being multiplied. If the decimal numbers also have units of measure attached to them, the units of measure are multiplied together just as was the case for fractions.

Example

1. Multiply 2.12 times 4.731.

$$\begin{array}{r} 2.12 \quad \text{(2 decimal places)} \\ \times\ 4.731 \ \text{(3 decimal places)} \\ \hline 10.02972 \ \text{(2 + 3 = 5 decimal places)} \end{array}$$

2. What is the area of a carpet that measures 4.2 ft by 3.5 ft?

$$4.2 \text{ ft} \times 3.5 \text{ ft} = 4.2 \times 3.5 \times \text{ft} \times \text{ft}$$

$$= 14.70 \text{ ft}^2 \text{ (square feet)}$$

3. What is the volume of a box that measures 2.4 cm by 1.58 cm by 10.2 cm?

$$2.4 \text{ cm} \times 1.58 \text{ cm} \times 10.2 \text{ cm} = 2.4 \times 1.58 \times 10.2 \times \text{cm} \times \text{cm} \times \text{cm}$$

$$= 38.6784 \text{ cm}^3 \text{ (cubic centimeters)}$$

Numbers in Exponential Form. For a review of the multiplication of numbers in exponential form, reread Appendix 1 in the textbook. Rewriting a number in exponential form so that it appears with a different power of 10 is just an application of

6

the rules for multiplication. For example, suppose we wanted to rewrite the number 4.86×10^5 as a number times 10^2. To do this, we would follow these steps:

1. Rewrite the original power of 10 as two powers of 10 multiplied times each other, with one of the two powers of 10 being the power that we want to end up with. In this example, we would write

$$10^5 = 10^{(3 + 2)} = 10^3 \times 10^2$$

2. Then multiply the coefficient by the power of 10 that we are not interested in, leaving behind the power of 10 that we wanted. In our example,

$$4.86 \times 10^5 = 4.86 \times 10^3 \times 10^2 = 4860 \times 10^2$$

Example

1. Rewrite the number 8.2×10^{-6} as a number times 10^{-4}.

$$8.2 \times 10^{-6} = 8.2 \times 10^{[-2 + (-4)]} = 8.2 \times 10^{-2} \times 10^{-4}$$
$$= 0.082 \times 10^{-4}$$

2. Rewrite the number 537.9×10^3 as a number whose coefficient is between 1 and 10.

Using the fact that $537.9 = 5.379 \times 10^2$, we have

$$537.9 \times 10^3 = 5.379 \times 10^2 \times 10^3 = 5.379 \times 10^{(2 + 3)}$$
$$= 5.379 \times 10^5$$

Self-Test

For questions 21 to 29, perform the indicated multiplication:

21. $\dfrac{3}{7} \times \dfrac{22}{12} \times \dfrac{4}{5}$

22. $60 \left(\dfrac{15}{21}\right) \left(\dfrac{16}{13}\right)$

23. $2.5 \text{ kg} \left(\dfrac{2.2 \text{ lb}}{1 \text{ kg}}\right)$

24. $\left(\dfrac{45 \text{ miles}}{1 \text{ hour}}\right) \left(\dfrac{1 \text{ hour}}{60 \text{ sec}}\right) \left(\dfrac{1.6 \text{ km}}{1 \text{ mile}}\right)$

25. $3.68 \text{ liter} \times 4.326 \text{ atm}$

26. $(12.4 \text{ m})(1.2 \text{ m})$

27. $(55 \times 10^4)(3.8 \times 10^2)$

28. $(0.043 \times 10^6)(0.02 \times 10^{-2})$

29. $(4.6 \times 10^{-8})(0.012 \times 10^{-3})$

For questions 30 to 34, rewrite the number in exponential form as indicated.

30. Rewrite 58.2×10^7 as a number times 10^5.

31. Rewrite 0.00356×10^{-4} as a number whose coefficient is between 1 and 10.

32. Rewrite 0.012×10^{-2} as a number times 10^{-4}.

33. Rewrite 46.267×10^{-3} as a number whose coefficient is between 1 and 10.

34. Rewrite 6.891×10^3 as a number times 10^6.

Division

The division of a number "a" by another number "b" can be expressed in several ways:

$$a \div b, \quad \frac{a}{b}, \quad a/b,$$

Fractions. Dividing a number by a fraction will give the same result as if the number were multiplied by the reciprocal, or inverse, of that fraction. The reciprocal of a fraction is formed by "flipping over" the fraction; for example, the reciprocal of $\frac{3}{5}$ is $\frac{5}{3}$. When numbers having units attached to them are divided, the units are expressed in the form of a "fraction" also.

Example

1. $15 \div \frac{5}{8} = 15 \times \frac{8}{5} = \frac{15 \times 8}{5} = \frac{120}{5} = 24$

2. $\dfrac{24}{\frac{2}{5}} = 24 \times \frac{5}{2} = \frac{24 \times 5}{2} = \frac{120}{2} = 60$

3. $\dfrac{75.6 \text{ g}}{14 \text{ mL}} = \frac{75.6}{14} \times \frac{g}{mL} = 5.4 \frac{g}{mL} = 5.4 \text{ g/mL}$

Numbers in Exponential Form. Reread Appendix 1 for a description of the procedure for dividing numbers expressed in exponential form.

For questions 35 to 40, perform the indicated division.

35. $5.796 \div 1.68$ *3.45*

36. $\dfrac{32}{1.28}$ *25*

37. $24 \div \dfrac{6}{7}$ *28*

38. $\dfrac{14}{\frac{5}{8}}$ *22⅘*

39. $\dfrac{478 \text{ miles}}{6.25 \text{ hr}}$ *76.48 mi/hr*

40. $\dfrac{0.275 \text{ mole}}{0.05 \text{ liter}}$ *5.5 mole/lite*

For questions 41 to 44, perform the indicated division and rewrite the answer as a number having a coefficient between 1 and 10.

41. $\dfrac{7196 \times 10^{12}}{2.57 \times 10^{25}}$ *$\dfrac{7,196^{16}}{2.57^{26}} = 2.8 \times 10^{-10}$*

43. $\dfrac{125 \times 10^{8}}{3125 \times 10^{2}}$ *$\dfrac{125^{11}}{3125000} = 4. \times 10^{4}$*

42. $\dfrac{37.5 \times 10^{2}}{0.025 \times 10^{-7}}$ *$\dfrac{37500}{2.5 - 08} = 1.5 \times 10^{2}$*

44. $\dfrac{3.2 \times 10^{2}}{1600 \times 10^{6}}$ *$\dfrac{3200}{1.6^{10}} = 2 \times 10^{-7}$*

Percent Notation

The term "percent" means "out of 100", and is denoted by the symbol "%". Thus, a statement such as "50% of the shirts are red" means that 50 out of 100 shirts (or half of the shirts) are red. This same proportion can also be written 50:100 or 50/100. To convert numbers between percentage notation and the other types of numbers we have discussed, use the following rules:

1. To write a decimal number as a percentage, move the decimal point 2 places to the right and add a percent sign. For example, 0.48 = 48%, 0.0732 = 7.32%, and 2.781 = 278.1%.

2. To write a fraction as a percentage, first convert the fraction to a decimal number, by dividing the denominator into the numerator, and then convert the decimal to a

percentage as described in Rule 1. For example,

$$\frac{3}{4} = 0.75 = 75\%$$

3. To write a percentage as a decimal number, move the decimal point 2 places to the left and remove the percent sign. For example, 71% = 0.71, 0.21% = 0.0021, and 130% = 1.3.

Self-Test

For questions 45 to 50, rewrite the given numbers as percents.

45. 0.46 _46%_   47. 0.045 _4.5%_   49. 2.671 _267.1%_

46. 4/5 _80%_   48. $\frac{2}{9}$ _22.2%_   50. 3.18 × 10⁻¹ _31.8%_

For questions 51 to 54, convert the percentages to decimals.

51. 4% _.04_   52. 20.5% _.205_   53. 7.67% _.0767_   54. 0.03% _.0003_

Solving Equations for an Unknown

An equation is a mathematical expression stating that two quantities are equal:

$$\frac{30}{36} = \frac{5}{6} \qquad \text{or} \qquad (12 \text{ eggs}) = (1 \text{ dozen eggs}).$$

Often we can state that two quantities are equal even when we don't know the exact value of some of the numbers in the equation. Such unknown values are represented by letters:

$$\frac{X}{4} = \frac{20}{8}, \qquad \text{or} \qquad P_1V_1 = P_2V_2$$

The process of figuring out the unknown value in an equation is called "solving the equation for the unknown". To solve an equation for an unknown quantity, there are only two rules to follow:

1. Try to undo what is being done to the unknown in the equation.

2. Always do the same thing to both sides of the equation, so that the two sides will remain equal.

1. Solve the following equation for the value of Z: $\frac{Z}{4} = \frac{20}{8}$.

    In this equation, Z is being divided by 4. To undo this we will have to multiply by 4—but we must remember to multiply both sides of the equation at the same time.

    $$4 \times \frac{Z}{4} = 4 \times \frac{20}{8}$$

    Or,

    $$Z = \frac{4 \times 20}{8} = 10$$

2. Solve the following equation for P_2: $P_1V_1 = P_2V_2$.

    We see that P_2 is being multiplied by V_2. To undo this, we will have to divide by V_2—again remembering to do the same thing to both sides of the equation.

    $$\frac{P_1V_1}{V_2} = \frac{P_2V_2}{V_2}$$

    Or,

    $$\frac{P_1V_1}{V_2} = P_2$$

Self-Test

For questions 55 to 58, solve the equation for X.

55. $\frac{X}{25} = \frac{760}{950}$

56. $16X = 86.4$

57. $\frac{275}{X} = \frac{5}{0.04}$

58. $\frac{X}{c+d} = \frac{ab}{Y}$

For questions 59 to 63, solve the equation for the indicated unknown.

59. $V_1M_1 = V_2M_2$; solve for M_2.

60. $K = °C + 273$; solve for $°C$.

61. $PV = nRT$; solve for V, and then solve for T.

62. $1.8(°C) = (°F) - 32$; solve for $°F$, and then solve for $°C$.

63. $\dfrac{P_1V_1}{T_1} = \dfrac{P_2V_2}{T_2}$; solve for P_2, then for V_2, then for T_2.

Answers to Self-Test Questions for Mathematics Review

1. 1.56×10^3 **2.** 7.432×10^6 **3.** 4.6×10^2 **4.** 1.45×10^2 **5.** 2.5×10^{-2} **6.** 7×10^{-7} **8.** 4.1×10^{-5} **9.** 2.5×10^4 **10.** 1.5×10^{-1}
11. 31.05 **12.** 0.1045 **13.** 340.1×10^6 **14.** 186.2315 **15.** 405×10^{-5} **16.** 144.864 **17.** 6.475 **18.** 120×10^5 **19.** 60×10^{-6} **20.** 1.5×10^{-8} **21.** $264/420 = 22/35$ **22.** $14{,}400/273 = 4800/91$ **23.** 5.5 lb
24. 1.2 km/sec **25.** 15.91968 liter-atm **26.** 14.88 m^2 **27.** 2.09×10^8 **28.** 8.6 **29.** 5.52×10^{-13} **30.** 5820×10^5 **31.** 3.56×10^{-7}
32. 1.2×10^{-4} **33.** 4.6267×10^{-2} **34.** 0.006891×10^6 **35.** 3.45
36. 25 **37.** 28 **38.** 22.4 **39.** 76.48 miles/hr **40.** 5.5 moles/liter
41. 2.8×10^{-10} **42.** 1.5×10^{12} **43.** 4×10^4 **44.** 2×10^{-7} **45.** 46%
46. 80% **47.** 4.5% **48.** 22% **49.** 267.1% **50.** $31/8\%$ **51.** 0.04
52. 0.205 **53.** 0.0767 **54.** 0.0003 **55.** $X = 20$ **56.** $X = 5.4$
57. $X = 2.2$

58. $X = \dfrac{ab(c+d)}{Y}$ **59.** $M_2 = \dfrac{V_1M_1}{V_2}$ **60.** $^\circ C = K - 273$

61. $V = \dfrac{nRT}{P}$; $T = \dfrac{PV}{nR}$ **62.** $^\circ F = 1.8\,^\circ C + 32$; $^\circ C = \dfrac{^\circ F - 32}{1.8}$

63. $P_2 = \dfrac{P_1V_1T_2}{T_1V_2}$; $V_2 = \dfrac{P_1V_1T_2}{T_1P_2}$; $T_2 = \dfrac{P_2V_2T_1}{P_1V_1}$

Chapter 1   THE STRUCTURE AND PROPERTIES OF MATTER

Chapters 1 through 4 introduce the basic chemical vocabulary that is used throughout the textbook. So, it is vital that you master the Important Terms within these chapters. If you have previously had a chemistry course, much of this will be review; you can check your knowledge by answering the self-test questions at the end of each section. If you have not had chemistry before, you will want to study these chapters slowly, becoming thoroughly familiar with each section before you go on. A little extra time spent studying these chapters now will make later chapters much easier to master.

1.1   What is Matter?

Matter has mass and occupies space. The amount of matter in an object determines the mass of the object. Mass is a measure of the resistance of an object to a change in its speed or in the direction of its motion. Weight is a measure of the gravitational attraction on an object. The mass of an object does not depend upon gravitational attraction, so it never changes.

Important Terms

matter   mass   weight

Self-Test

For questions 1 to 3, answer true (T) or false (F).

1. The mass of an object does not vary with a change in location.

2. A Volkswagen has a greater mass than a Cadillac.

3. If a stone weighs 20 pounds on the moon, it would weigh about 15 pounds on Earth.

1.2   Composition of Matter

Atoms are particles that make up matter. They are extremely small, and yet have been shown to be composed of many smaller particles. The most important subatomic particles are the proton, the neutron, and the electron.

Important Terms

atom   proton   neutron   electron

1.3   Classes of Matter: Elements, Compounds, and Mixtures

Elements are the simplest substances encountered in the chemical laboratory.   They make up all matter, either alone or in combination with other elements.   Elements are pure substances that contain only one kind of atom.   Compounds, by contrast, are pure substances composed of atoms of more than one element.   A molecule is a particle composed of two or more atoms joined together.

Compounds, no matter how they are formed, always contain the same type of atoms in a definite ratio or proportion by weight.   This is known as the Law of Definite Proportions.   Mixtures, on the other hand, contain two or more substances in any proportion (such as salt in water or carbon monoxide in exhaust fumes). When substances combine to form a compound, a new substance with different properties is formed, but substances in a mixture keep their individual properties.   Mixtures can be identified as either homogeneous or heterogeneous.

Important Terms

element	compound	atom
molecule	mixture	homogeneous
heterogeneous	Law of Definite Proportions	

Self-Test _____

For questions 4 to 8, answer true (T) or false (F).

4. When heated in a laboratory, an element can break down to form two or more simpler substances.

5. A molecule is composed of two or more atoms.

6. Compounds contain atoms from two or more different elements.

7. Substances contained in a mixture can be identified by their individual properties.

8. An atom is the smallest particle of an element that keeps the properties of that element.

14

9. Identify each of the following as a homogeneous mixture (hom) or heterogeneous mixture (het).

het (a) concrete   *het* (c) salad dressing   *hom* (e) honey   *het* (g) sand
hom (b) gasoline   (d) black coffee   (f) cytoplasm
       Hom   *Het*

1.4   Names and Symbols for the Elements

An element can be represented by its symbol, which in most cases is the first letter of the element's name (for example, N for nitrogen and P for phosphorus). If two letters are used, the second letter is not capitalized (for example, Mg for magnesium and He for helium).

Self-Test _____

Use the table on the inside back cover of the textbook to help you with questions 10 and 11.

10. Write the symbol for each of the following elements.

    (a) hydrogen   H   (e) potassium   K   (h) copper   Cu
    (b) carbon   C   (f) manganese   MA   (i) chlorine   Cl
    (c) oxygen   O   (g) iron   Fe   (j) zinc   Zn
    (d) sodium   Na

11. Write the names of each of the following elements.

    (a) N   nitroge   (e) S   Sulfur   (h) Co   Cobalt
    (b) F   Iron   (f) Ca   Carbien   (i) Se   Selenium
    (c) Mg   Magnesium   (g) Cr   Chromium   (j) Sn   Tin
    (d) P   Phosphorus

1.5   The Three States of Matter

Matter exists in three states, which can be converted one to another by adding or removing energy. These three states are solid, liquid, and gas.

Important Terms

     solid   liquid   gas

1.6   Physical and Chemical Changes

When a physical change occurs, the nature of the substance remains the same--only its form is changed. In a chemical change, new substances with different chemical and physical characteristics are formed.

Important Terms

physical change   chemical change

Self-Test _____

12. Identify each of the following as either a chemical change (C) or a physical change (P).

   P (a) Alcohol boiling
   C (b) Antacids neutralizing stomach acid
   P (c) Pulverizing an aspirin
   P (d) Dissolving an aspirin in water
   C (e) Wood burning
   C/P (f) Sugar being used by a cell to produce energy, carbon dioxide, and water
   C/P (g) Carbon monoxide reacting with oxygen to form carbon dioxide
   P (h) The sun drying swimmers on the beach

1.7   Scientific Method

The scientific method is a precise way of studying the world around us through observation, development of a hypothesis, testing that hypothesis, and then proposing a scientific theory.

Important Term

scientific method

1.8   Accuracy and Precision

To be useful, data collected to test a hypothesis must be both accurate and precise. A precise measurement is one that is reproducible. An accurate measurement is one that comes close to the true value.

Important Terms

accurate   precise

1.9   Significant Figures

Significant figures, or significant digits, indicate the precision with which a measurement was made. The correct number of significant figures to use in a measurement is the number of digits that are certain plus the first digit for which the measurement is uncertain.

Important Terms

significant figures   significant digits

Self-Test ────────────────────────────────────

13. A certain volume of water (37.00 mL) was measured in a 50 mL beaker, a 50 mL graduated cylinder, and a 50 mL buret by three students. Their recorded data was as follows:

		Student A	Student B	Student C
I.	50 mL beaker	40 mL	40 mL	40 mL
II.	50 mL graduated cylinder	37.0 mL	38.2 mL	36.5 mL
III.	50 mL buret	37.05 mL	37.10 mL	36.95 mL

(a) Which set of measurements is the most precise: I, II, or III?

(b) Which set of measurements is the most accurate?

(c) How many significant figures are there in each of the measurements in III?

14. Give the number of significant figures in each of the following measurements:

(a) 225 lb
(b) 22.50 g
(c) 100 m
(d) 0.207 liter
(e) 0.0036 mg
(f) 25.068 kg

1.10 Unit Factor Method

Unit factors, or conversion factors, are ratios that are equivalent to the number "one". They are very important in mathematical calculations in this textbook. Unit factors are used to change the units that are stated in a problem to the units that you want in the answer. A unit factor or conversion factor can be formed from any equality by dividing both sides of the equation by one of the sides. For example, from the equality 1 in. = 2.54 cm, we can create two unit factors:

$$\frac{1 \text{ in.}}{2.54 \text{ cm}} \quad \text{and} \quad \frac{2.54 \text{ cm}}{1 \text{ in.}}$$

unit factor   conversion factor

1.11 The SI System of Units

The SI System of Units is used by scientists throughout the
world. The SI system is identical to the metric system, but it
uses different units for reference. For example, in the metric
system the reference unit of volume is the liter and in the SI
system it is the cubic meter. The SI and metric systems have an
advantage over the English system in that all units of measure
are related to their subparts by multiples of ten. The names of
larger or smaller units are derived from the reference unit by
using the appropriate prefix. The following are the basic units
in the SI system, along with some other commonly used units.

Measurement	SI System	Common Units
Length	meter (m)	kilometer (km) centimeter (cm) millimeter (mm)
Mass	kilogram (kg)	gram (g) milligram (mg) microgram (μg)
Volume	cubic meter (m^3)	liter (L) milliliter (mL) cubic centimeter (cc)

Most conversions between units in the SI system and units in the
English system can be done using the conversion factors given on
the inside back cover of your text.

Important Terms

International System of Units   metric system
    (SI system)   English system

Self-Test _____

15. Complete the following table.

	Prefix	Multiple
(a)	kilo-	( )
(b)	centi-	( )
(c)	( )	10
(d)	milli-	( )
(e)	hecto-	( )
(f)	( )	0.1
(g)	( )	0.000001

18

16. For each of the following groups, place the units in order from the smallest to the largest.

    (a) centigram, microgram, kilogram, milligram
    (b) deciliter, milliliter, liter, centiliter
    (c) meter, micrometer, centimeter, millimeter

17. Write two unit factors for each of the following relationships.

    (a) 1 dozen eggs contains 12 eggs.
    (b) 1000 milliliters = 1 liter.
    (c) There are 3 feet in a yard.
    (d) 1 centimeter = 0.39 inch.

1.12 Length

The unit of length in the SI system is the meter, which is slightly longer than a yard (1 meter = 3.28 feet).

Important Terms

    length   meter

Example _____

1. How many centimeters are there in 2.7 meters?

    The relationship between centimeters and meters is

$$1 \text{ meter (m)} = 100 \text{ centimeters (cm)}.$$

    From this equation we can write two unit factors:

$$\frac{1 \text{ m}}{100 \text{ cm}} \quad \text{and} \quad \frac{100 \text{ cm}}{1 \text{ m}}$$

    Our problem asks: 2.7 meters = ? centimeters; therefore, the second unit factor is the one to use.

$$2.7 \text{ m} \times \frac{100 \text{ cm}}{1 \text{ m}} = 270 \text{ cm}$$

2. How many millimeters long is a 1.5 inch incision?

    The solution to this problem requires two relationships.

$$10 \text{ millimeters (mm)} = 1 \text{ centimeter (cm)}$$

Unit factors: $\dfrac{10 \text{ mm}}{1 \text{ cm}}$ or $\dfrac{1 \text{ cm}}{10 \text{ mm}}$

and: 2.54 cm = 1 inch (in.)

Unit factors: $\dfrac{2.54 \text{ cm}}{1 \text{ in.}}$ or $\dfrac{1 \text{ in.}}{2.54 \text{ cm}}$

Our problem asks: 1.5 inch = ? millimeters; therefore, we choose the unit factors necessary to produce the answer in the correct units:

$$1.5 \text{ in.} \times \frac{2.54 \text{ cm}}{1 \text{ in.}} \times \frac{10 \text{ mm}}{1 \text{ cm}} = 38 \text{ mm}$$

Self-Test

For questions 18 to 25, complete the conversions required.

18. 3.6 km = _____ m

19. 125 cm = _____ m

20. 0.25 m = _____ mm

21. 468 m = _____ km

22. 37.0 ft = _____ m

23. 100 yd = _____ m

24. 40 mi = _____ km

25. 65 mm = _____ in.

1.13 Mass

The unit of mass in the SI system is the kilogram, which equals 2.2 pounds. The unit of mass in the metric system is the gram.

Important Terms

mass   gram   kilogram

Example

1. How many milligrams are there in a 0.18 gram sample of radioactive tracer?

The relationship between grams and milligrams is

1 gram (g) = 1000 milligrams (mg)

Unit factors: $\dfrac{1 \text{ g}}{1000 \text{ mg}}$ or $\dfrac{1000 \text{ mg}}{1 \text{ g}}$

Our problem asks: 0.18 g = ? mg; therefore, the second unit factor is the one to use.

$$0.18 \text{ g} \times \frac{1000 \text{ mg}}{1 \text{ g}} = 180 \text{ mg}$$

2. The weight limit for luggage on an international flight is 40 pounds, but the scale in the hotel is calibrated in kilograms. How many kilograms are there in 40 pounds?

The relationship between pounds and kilograms is

$$1 \text{ kilogram (kg)} = 2.2 \text{ pounds (lb)}$$

Unit factors: $\dfrac{1 \text{ kg}}{2.2 \text{ lb}}$ or $\dfrac{2.2 \text{ lb}}{1 \text{ kg}}$

Our problem asks: 40 lb = ? kg; therefore, the first unit factor is the one to use.

$$40 \text{ lb} \times \frac{1 \text{ kg}}{2.2 \text{ lb}} = 18 \text{ kg}$$

Self-Test _____

For questions 26 to 35, complete the conversions required.

26. 0.4 g = _____ mg	31. 1.50 lb = _____ g
27. 2.4 kg = _____ g	32. 2.1 kg = _____ lb
28. 45 mg = _____ g	33. 30.0 oz = _____ kg
29. 8.7 g = _____ kg	34. 0.700 kg = _____ oz
30. 0.02 g = _____ μg	35. 6.80 oz = _____ g

1.14 Volume

The SI unit of volume is the cubic meter, a fairly large unit of measure. The liter (1000 liters = 1 cubic meter) is a more commonly used unit of volume. One liter is slightly larger than a quart.

Important Terms

volume	liter	cubic meter
milliliter	cubic centimeter	

1. How many 5-milliliter doses are there in a bottle containing 0.3 liter of ampicillin?

    This problem can be solved in two steps. First, we answer the question 0.3 liter = ? milliliters. The relationship between liters and milliliters is

    $$1 \text{ liter} = 1000 \text{ milliliters}$$

    Unit factors: $\dfrac{1 \text{ liter}}{1000 \text{ mL}}$ and $\dfrac{1000 \text{ mL}}{1 \text{ liter}}$

    Therefore, 0.3 liter $\times$ $\dfrac{1000 \text{ mL}}{1 \text{ liter}}$ = 300 mL

    Second, we must answer the question 300 mL = ? doses. The relationship between doses and milliliters is

    $$1 \text{ dose} = 5 \text{ mL}$$

    Unit factors: $\dfrac{1 \text{ dose}}{5 \text{ mL}}$ and $\dfrac{5 \text{ mL}}{1 \text{ dose}}$

    Therefore, 300 mL $\times$ $\dfrac{1 \text{ dose}}{5 \text{ mL}}$ = 60 doses

2. How many cups are there in 0.60 liter?

    The two relationships needed to solve this problem are

    1 liter = 1.06 quarts: $\dfrac{1 \text{ liter}}{1.06 \text{ qt}}$ or $\dfrac{1.06 \text{ qt}}{1 \text{ liter}}$
    and

    1 quart = 4 cups: $\dfrac{1 \text{ qt}}{4 \text{ cups}}$ or $\dfrac{4 \text{ cups}}{1 \text{ qt}}$

    Using a unit factor from each relationship, we can solve the problem in one step.

    0.60 liter $\times$ $\dfrac{1.06 \text{ qt}}{1 \text{ liter}}$ $\times$ $\dfrac{4 \text{ cups}}{1 \text{ qt}}$ = 2.5 cups

Self-Test ————————————————————————————

For questions 36 to 45, complete the conversions required.

36. 250 dL   = _____ liter   41. 9.0 gal = _____ liter

37. 150 mL   = _____ cL   42. 2.50 pt = _____ mL

38. 2.2 liter = _____ mL   43. 405 mL   = _____ qt

39. 425 mL   = _____ liter   44. 250 mL   = _____ cup

40. 3.4 cL   = _____ liter   45. 0.50 cup = _____ liter

46. The doctor prescribes a dose of 1.5 teaspoons of penicillin three times a day until the bottle is empty. The bottle contains 250 mL of penicillin. For how many days will you be giving penicillin to your son? (1 teaspoon = 5.0 mL)

1.15 Temperature

There are three temperature scales used for measuring temperature. You are probably most familiar with the Fahrenheit scale. The Celsius scale is used in the metric system, and the Kelvin scale is used in the SI system. A one degree temperature difference is the same in the Celsius and Kelvin scales, but the scales start at different reference points. The freezing point of water is 0°C and 273.15 K. Zero degrees on the Kelvin scale corresponds to absolute zero, the lowest temperature it is theoretically possible to reach.

Important Terms

Fahrenheit scale   Kelvin scale
Celsius scale   absolute zero

Example _____

1. A child has a temperature of 104°F. What temperature is this on the Celsius scale?

    To convert between the Fahrenheit and the Celsius scales, we use the following relationship:

    $$°C = \frac{5}{9} (°F - 32)$$

    Substituting in this equation, we have

    $$°C = \frac{5}{9} (104 - 32) = 40°C$$

2. Neon boils at 27 K. What temperature is this on the Celsius scale?

To convert between the Kelvin and the Celsius scales, we use
the following relationship: K = °C + 273 (actually 273.15,
but for solving problems in this chapter we will use the
conversion factor rounded to the nearest whole number).
Substituting in this relationship, we have

$$27 = °C + 273$$

$$°C = 27 - 273 = -246°C$$

Self-Test

47. Perform the following conversions:

(a) 40°C = ____ K   (c) 100 K = ____ °C   (e) 253 K = ____ °F

(b) 77°F = ____ K   (d) 110°C = ____ °F   (f) 23°F = ____ °C

1.16 Density and Specific Gravity

The density of a substance is the mass of that substance per unit
of volume. The specific gravity of a substance compares the
density of that substance to the density of water (density of
water = 1 g/cc or 1 g/mL at 4°C).

$$density = \frac{mass}{Volume} \qquad specific\ gravity = \frac{density\ of\ sample}{density\ of\ water}$$

Important Terms

density   specific gravity

Example

1.   A rock has a mass of 70 g. When it is placed in a 100 mL
     graduated cylinder containing water, the water level rises
     from the 41 mL mark to the 59 mL mark. What is the density
     of the rock?

     The volume of the rock is equal to the volume of the water
     displaced by the rock. Therefore,

     $$volume = 59\ mL - 41\ mL = 18\ mL = 18\ cc$$

     $$density\ of\ the\ rock = \frac{70\ g}{18\ cc} = 3.9\ g/cc$$

2. An oil sample has a density of 0.82 g/cc. What is the specific gravity of the oil?

$$\text{specific gravity} = \frac{0.82 \text{ g/cc}}{1.00 \text{ g/cc}} = 0.82$$

3. Alcohol has a density of 0.79 g/cc. What is the weight of a 7.5 mL sample?

We can use the value of the density as a unit factor to solve this problem.

$$7.5 \text{ mL} \times \frac{0.79 \text{ g}}{1 \text{ mL}} = 5.9 \text{ g}$$

Self-Test

48. What is the density of a solid that has a volume of 23 mL and a mass of 163 g?

49. What is the specific gravity of a liquid that has a density of 0.950 g/cc? What is the mass of a 250-mL sample of this liquid?

50. A 75.0-mL sample of urine weighs 86.25 g. What is the specific gravity of this urine sample?

51. What is the volume of a solution that has a mass of 125 g and a density of 1.40 g/cc?

52. What is the density of a 25-g sample of solid if the water level rises from the 34 mL mark to the 47 mL mark when the sample is completely emersed in a graduated cylinder containing water?

Review Self-Test

53. A box measures 4.00 inches on each side. What is its volume in cubic centimeters?

54. What is the length in centimeters of a five-inch incision?

55. If a circular birthmark has a diameter of 0.25 inch, what is its radius in millimeters?

56. A patient is 5 feet 2 inches tall. What is her height in meters?

57. A sleeping pill contains 100 mg of powder. How many grams of powder is this?

58. What is the weight in kilograms of a 140-pound patient?

59. If the recommended dose of a drug is 4 milliliters, how many doses will there be in a bottle containing 0.2 liter?

60. How many grams of drug are there in 4 ampules if each ampule contains 7.5 grains (1 grain = 60 milligrams)?

61. A vial of Kanamycin is labeled 0.25 g/mL. The required dose of the drug is 15 mg for each kilogram the patient weighs. If a patient weighs 55 pounds, how many milliliters are necessary for the required dose?

62. If a physician ordered 2 grams of drug to be given in four equal doses and the drug were available in 250-mg capsules, how many capsules would you give for each of the four doses?

1. T **2.** F **3.** F **4.** F **5.** T **6.** T **7.** T **8.** T **9.**(a) het (b) hom (c) het (d) hom (e) hom (f) het (g) het **10.**(a) H (b) C (c) O (d) Na (e) K (f) Mn (g) Fe (h) Cu (i) Cl (j) Zn **11.**(a) nitrogen (b) fluorine (c) magnesium (d) phosphorus (e) sulfur (f) calcium (g) chromium (h) cobalt (i) selenium (j) tin **12.**(a) P (b) C (c) P (d) P (e) C (f) C (g) C (h) P **13.**(a) I (b) III (c) 4 **14.**(a) 3 (b) 4 (c) 1 (d) 3 (e) 2 (f) 5 **15.**(a) 1000 (b) 0.01 (c) deka- (d) 0.001 (e) 100 (f) deci- (g) micro **16.** (a) microgram, milligram, centigram, kilogram (b) milliliter, centiliter, deciliter, liter (c) micrometer, millimeter, centimeter, meter

17. (a) $\dfrac{1\ dozen}{12\ eggs}$, $\dfrac{12\ eggs}{1\ dozen}$ (b) $\dfrac{1000\ mL}{1\ liter}$, $\dfrac{1\ liter}{1000\ mL}$

(c) $\dfrac{3\ ft}{1\ yd}$, $\dfrac{1\ yd}{3\ ft}$ (d) $\dfrac{1\ cm}{0.39\ in.}$, $\dfrac{0.39\ in.}{1\ cm}$

18. 3600 m **19.** 1.25 m **20.** 250 mm **21.** 0.468 km (Watch the significant figures in your answers.) **22.** 11.3 m **23.** 91.5 m **24.** 64 km **25.** 2.6 in. **26.** 400 mg **27.** 2400 g **28.** 0.045 g **29.** 0.0087 kg **30.** 20,000 μg **31.** 682 g **32.** 4.6 lb **33.** 0.849 kg **34.** 24.7 oz **35.** 192 g **36.** 25 liter **37.** 15 cl **38.** 2200 mL **39.** 0.425 liter **40.** 0.034 liter **41.** 34 liter **42.** 1180 mL **43.** 0.429 qt **44.** 1.06 cup **45.** 0.12 liter

46. $\dfrac{1.5\ tsp}{1\ dose} \times \dfrac{5.0\ mL}{1\ tsp} \times \dfrac{3\ doses}{1\ day} = \dfrac{22.5\ mL}{1\ day}$ 250 mL $\times \dfrac{1\ day}{22.5\ mL} = 11$ days

47.(a) 313 K (b) 298 K (c) −173°C (d) 230°F (e) −4°F (f) −5°C **48.** 7.1 g/cc **49.** 0.950, 238 g **50.** 1.15 **51.** 89.3 cc **52.** 1.9 g/cc **53.** 1050 cm^3 **54.** 12.7 cm **55.** 3.2 mm **56.** 1.6 m **57.** 0.1 g **58.** 63.6 kg **59.** 50 doses

60. 7.5 grains $\times \dfrac{60\ mg}{1\ grain} \times \dfrac{1\ g}{1000\ mg} = 1.8$ g

61. 55 lb $\times \dfrac{1\ kg}{2.2\ lb} \times \dfrac{15\ mg}{1\ kg} \times \dfrac{1\ g}{1000\ mg} \times \dfrac{1\ mL}{0.25\ g} = 1.5$ mL

62. 2 capsules/dose

Chapter 2   ENERGY

In this chapter we define energy and look at different types of energy and the characteristics of each type of energy.

2.1   What is Energy?

Energy is the ability or capacity to cause various kinds of change. For example, pedaling a bicycle up a hill requires your muscles to expend energy to move the bicycle from one place to another.

Important Term

     energy

2.2   Kinetic Energy

Kinetic energy is the energy an object possesses as a result of its motion. The amount of kinetic energy possessed by an object depends upon the mass of the object and the speed with which it is moving (its velocity). Because the particles that make up matter are in constant motion, they possess kinetic energy. The temperature of a substance is a measure of the average kinetic energy of all the particles that make up that substance. The higher the temperature of a substance, the greater the kinetic energy of its particles.

Important Terms

     kinetic energy   temperature

2.3   Potential Energy

Potential energy is stored energy. A substance can possess potential energy by virtue of its position. A truck parked at the top of a hill possesses potential energy that would be converted to kinetic energy if its brakes were to fail, causing it to roll down the hill. A hibernating animal contains potential energy in its stored body fat; this energy is slowly released to sustain the animal over the winter.

Important Term

     potential energy

1. Indicate whether each of the following possesses kinetic energy (KE) or potential energy (PE):
    (a) Wood
    (b) A moving freight train
    (c) Snow in the mountains
    (d) Propane gas
    (e) The Mississippi River
    (f) An arrow in a drawn bow
    (g) A football player just before the ball is snapped
    (h) A football player just after the ball is snapped

2. List the following in order of increasing kinetic energy:
    (a) water at 25°C   (c) water at 325 K   (e) water at 240°F
    (b) water at 28°F   (d) water at -75°C   (f) water at 140 K

2.4 Heat Energy

Heat is energy that is transferred from one place to another because of a difference in temperature. Heat is measured in units called calories. A calorie is the amount of heat energy required to raise the temperature of one gram of water one degree Celsius. In the SI system, the unit of heat energy is the joule (1 cal = 4.184 J).

The specific heat of a substance is the amount of energy required to raise the temperature of one gram of a substance one degree Celsius. The amount of heat necessary to change the temperature of a sample of a substance depends upon the specific heat of the substance, the mass of the sample, and the change in the temperature.

$$\text{calories} = g \times \Delta t \times \frac{1 \text{ cal}}{g \ °C}$$

A food Calorie is equivalent to 1000 calories, or one kilocalorie. The number of kilocalories of energy in a sample of food can be measured a calorimeter. The basal metabolism rate is the minimum amount of energy required daily to maintain a human body at rest.

Important Terms

heat	calorie
kilocalorie	joule
specific heat	calorimeter
basal metabolism rate (BMR)	

When a 6-gram sample of butter is burned in a calorimeter, the temperature of the 500 mL of water in the calorimeter increases from 27 to 63°C. How many kilocalories are contained in one gram of butter? (Remember that the density of water = 1 g/mL.)

To solve this problem we use the equation,

$$cal = g \text{ of water} \times \Delta t \times \frac{1 \text{ cal}}{g \text{ °C}}$$

Therefore,

$$cal = 500 \text{ g} \times (63°C - 27°C) \times \frac{1 \text{ cal}}{g \text{ °C}}$$

$$= \frac{500 \text{ g} \times 36°C \times 1 \text{ cal}}{g \text{ °C}}$$

$$= 18,000 \text{ cal or } 18 \text{ kcal}$$

If 6 grams of butter released 18 kilocalories of energy then one gram of butter would contain 18/6 or 3 kilocalories.

Self-Test

3. Complete the following conversions:
    (a) 250 cal = _____ kcal   (c) 25 Calories = _____ kcal
    (b) 3.4 kcal = _____ cal   (d) 9500 joule = _____ kcal

4. How many kilocalories are contained in one gram of hamburger if the energy released by a 10-gram sample of hamburger raised the temperature of 700 grams of water in a calorimeter from 24 to 64°C?

2.5   Changes in State

Matter exists in three states: solid, liquid, and gas. A change of state occurs when matter goes from one state to another. Changes in state can require energy as when ice melts or can give off energy as when gas condenses. Changes that require energy are called endothermic and those that give off energy are called exothermic.

Important Terms

endothermic   exothermic

2.6 Electromagnetic Energy

The visible light that we see is only a tiny part of the entire range of energy called electromagnetic energy or electromagnetic radiation. This energy travels in waves. Electromagnetic radiation with long wavelengths (for example, radiowaves) has low energy, but electromagnetic radiation with short wavelengths (for example, gamma rays) has extremely high energy. Electromagnetic radiation has different uses and different effects on living organisms depending upon its energy.

Important Terms

electromagnetic spectrum   wavelength
microwave radiation   visible light
ultraviolet radiation   gamma rays
infrared radiation   X rays

Self-Test _____

For questions 5 to 13, choose the best answer.

5. Man-made high energy radiation is
    (a) gamma rays   (c) radio waves
    (b) X rays   (d) infrared radiation

6. The radiation with the lowest energy is
    (a) radio waves   (c) microwave radiation
    (b) gamma rays   (d) infrared radiation

7. This radiation causes sunburn.
    (a) gamma rays   (c) ultraviolet radiation
    (b) red light   (d) infrared radiation

8. This high energy radiation is produced by natural sources.
    (a) radio waves   (c) microwave radiation
    (b) gamma rays   (d) X rays

9. The radiation with the longest wavelength is
    (a) radio waves   (c) microwave radiation
    (b) gamma rays   (d) X rays

10. This visible light has the highest energy.
    (a) red light   (c) yellow light
    (b) blue light   (d) purple light

11. Highly penetrating radiation is
    (a) radio waves   (c) microwave radiation
    (b) gamma rays   (d) infrared radiation

12. This radiation is given off by warm objects
    (a) X rays
    (b) radio waves
    (c) ultraviolet radiation
    (d) infrared radiation

13. This electromagnetic radiation is used for cooking.
    (a) gamma rays
    (b) X rays
    (c) microwave radiation
    (d) infrared radiation

2.7 Law of Conservation of Energy (optional section)

In any process, energy must be conserved. This means that the total amount of energy at the end of a process must equal the total amount of energy at the beginning of the process. Energy can neither be created nor destroyed; it can only change form. Each of these statements is a way of saying the Law of Conservation of Energy, or the First Law of Thermodynamics.

For example, the potential energy stored in a log does not disappear when the log is burned, but rather is converted into heat and light energy. Most of the light energy that reaches the earth from the sun is converted into heat energy, but a small amount is absorbed by green plants and converted into chemical energy.

Important Terms

    Law of Conservation of Energy
    First Law of Thermodynamics

2.8 Entropy (optional section)

All naturally occurring processes tend toward a state of lower energy and higher disorder or randomness. The term given to this disorder or randomness is entropy. The more disordered a system, the greater is its entropy. The Second Law of Thermodynamics states that the entropy or disorder of the universe is increasing.

In order for nonspontaneous reactions (those reactions that go toward higher energy and lower entropy) to occur, they must be coupled with reactions that produce energy. For every system of such coupled reactions, we find that the total entropy of the system increases.

A living organism contains highly complex and organized structures and is, therefore, in a state of very low entropy. The natural tendency is for these structures to break down, so that the living organism must constantly expend energy from food that it consumes to maintain its state of low entropy. When the organism dies and can no longer generate this energy, the natural process toward increased entropy (that is, decay) begins.

Important Terms

entropy
Second Law of Thermodynamics

spontaneous reaction
nonspontaneous reaction

Self-Test _____

For questions 14 to 17, fill in the blank with the correct word or words.

14. The word that describes the disorder of a system is _____.

15. All naturally occurring processes tend toward a state of _____ energy and _____ entropy.

16. When a process occurs, the energy at the end of the process must _____ the energy at the beginning of the process.

17. The randomness of the universe is _____ .

18. Which one in each of the following pairs has the higher entropy?
     (a) A new deck of cards, or a shuffled deck of cards
     (b) Water as a liquid, or water as a gas
     (c) A sugar cube, or sugar dissolved in a cup of coffee
     (d) Compost, or the grass in a lawn

Answers to Self-Test Questions in Chapter 2

1. (a) PE (b) KE (c) PE (d) PE (e) KE (f) PE (g) PE (h) KE
2. f,d,b,a,c,e **3.** (a) 0.250 kcal (b) 3400 cal (c) 25 kcal
(d) 2.3 kcal
4. 2.8 kcal

$$700 \text{ g} \times (64°C - 24°C) \times \frac{1 \text{ cal}}{\text{g °C}} = 28,000 \text{ cal} = 28 \text{ kcal}$$

28 kcal/10 g = 2.8 kcal/1 g

5. b **6.** a **7.** c **8.** b **9.** a **10.** d **11.** b **12.** d **13.** c
14. entropy **15.** lower, higher **16.** equal **17.** increasing
18. (a) shuffled cards (b) water as a gas (c) dissolved sugar
(d) compost

Chapter 3   ATOMIC STRUCTURE

In this chapter, we shall study the structure of the atom—the smallest unit of an element that has the properties of that element. There may be some vocabulary that is unfamiliar to you, so be sure that you understand the definitions of all the Important Terms.

3.1   The Parts of the Atom

The atom is not indivisible, but rather is made up of many particles, the most important of which are protons, neutrons, and electrons. The positive protons and the neutral neutrons are found in the small dense nucleus at the center of the atom. The negative electrons are found in a relatively large region surrounding the nucleus.

Important Terms

    proton   neutron   electron   nucleus

3.2   Atomic Number and Mass Number

The atomic number of an atom is equal to the number of protons in the nucleus of the atom. It is this number of protons that determines which element the atom represents. Each element has a specific and different atomic number. The number of electrons in a neutral atom is also equal to the atomic number. For example, the atomic number of sulfur is 16. Therefore, a neutral atom of sulfur will contain 16 protons and 16 electrons.

The mass number of an atom equals the number of protons plus the number of neutrons in its nucleus. Protons and neutrons have extremely small masses. Even so, they make up most of the mass of the atom since the mass of the electron is only 0.0005 times that of the proton or neutron. There are several shorthand ways to indicate the mass number and atomic number of an atom.

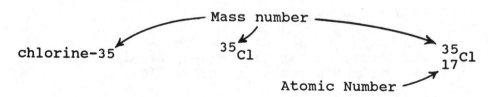

chlorine-35 ^{35}Cl $^{35}_{17}Cl$

Important Terms

atomic number   mass number

Self-Test _____

For questions 1 to 6, fill in the blank with the appropriate word or words.

1.   The nucleus of an atom contains _____ and _____ .

2.   The identity of an atom is determined by the number of _____ in its nucleus.

3.   To determine the number of electrons in a neutral atom, you need to know the _____ .

4.   The _____ of an atom equals the sum of the neutrons and protons in the atom's nucleus.

5.   The mass of a proton is greater than the mass of a _____ and is the same as the mass of a _____ .

6.   The electron has a charge of _____, the proton a charge of _____, and the neutron a charge of _____.

7.   Give the number of protons, neutrons, and electrons in a neutral atom of each of the following:

   (a) hydrogen-2   (d) ^{28}Si

   (b) carbon-13   (e) $^{79}_{35}$Br

   (c) oxygen-18

3.3   Isotopes

In many cases all the atoms of one element are not alike. They must have the same number of protons, but they can differ in the number of neutrons in their nuclei. Atoms of an element that have different numbers of neutrons in their nuclei are called isotopes.

Important Term

isotope

3.4 Atomic Weight

Since atoms are so small, an arbitrary scale of relative weights has been established to measure them. In this scale, an atom of carbon-12 has been assigned a mass of exactly 12 atomic mass units (amu). Any sample of an element we might obtain from the world around us will contain a mixture of the isotopes of that element. As a result, the observed weight of that element is based on the relative abundance of the isotopes of that element. The atomic weight of an element is calculated by taking a weighted average of the masses of the isotopes of that element.

Important Terms

    atomic weight   weighted average
    atomic mass unit (amu)

Example _____

Bromine has two isotopes: bromine-79 and bromine-81. The relative abundance of bromine-79 is 50.54% and that of bromine-81 is 49.46%. To calculate the atomic weight of bromine, we multiply the mass of each isotope by its percent abundance and then the sum the products.

$$
\begin{aligned}
79 \times 50.54\% &= 79 \times 0.5054 = 39.93 \\
81 \times 49.46\% &= 81 \times 0.4946 = \underline{40.06} \\
& 79.99 \quad \text{or 80 amu}
\end{aligned}
$$

Self-Test _____

For 8 to 11, identify the statements as true (T) or false (F):

8. Isotopes of an element differ in the number of protons in their nuclei.

9. The actual mass of a carbon-12 atom is 12 grams.

10. The atomic weight of an element is a weighted average of the masses of its isotopes.

11. The atomic weight of an element with only one isotope equals the mass of the isotope.

12. Calculate the atomic weight of silicon to two decimal places. (See Table 3.2 in the text for the relative abundance of the isotopes of silicon.)

3.5   The Quantum Mechanical Model of the Atom

Since atoms are so very small, scientists have used experimental data to develop elaborate theoretical models of the structure of the atom. Niels Bohr's model, based on the experimental data of Ernest Rutherford, pictured the atom as a small dense nucleus surrounded by electrons. The electrons could not assume just any position around the nucleus, but rather occupied positions corresponding to specific energy levels. The energy of these levels increases as the electrons move farther from the nucleus. To move from one energy level to another, the electron must absorb or give off an amount of energy exactly equal to the difference in energy between the two energy levels. Each energy level is designated by a number, and can hold only a certain maximum number of electrons.

Because Bohr's model of the atom failed to explain some experimental data, a new model based on complicated mathematical calculations was developed. This model, called the quantum mechanical model, states that the position of the electron can never be determined exactly. Instead of specific orbits, the position of the electron is now described in terms of probability regions called orbitals. An orbital can contain zero, one, or two electrons.

Each energy level contains sublevels consisting of one or more atomic orbitals. The first energy level contains one orbital, the 1s orbital, whose probability distribution or shape is that of a sphere. The second energy level contains a 2s orbital and three 2p orbitals. The third energy level contains one 3s orbital, three 3p orbitals, and five 3d orbitals. The fourth energy level contains one 4s orbital, three 4p orbitals, five 4d orbitals, and seven 4f orbitals.

The rules for filling these orbitals are:

1.   An orbital can hold no more than two electrons, which must be spinning in different directions.

2.   An electron will not pair up with another electron if there is another orbital of the same energy available.

3.   The order of filling the orbitals is shown in Figure 3.4. Note that beginning with energy level 3 the energy levels overlap, so electrons will begin filling a higher energy level before the lower one is completely filled.

Important Terms

Bohr model   energy level
quantum mechanical model

37

3.6 Electron Configurations

The arrangement of electrons or electron configuration of an atom can be shown in several ways. One is to list the orbitals in order of increasing energy and to use a superscript to indicate the number of electrons in each orbital. A second method called the orbital diagram uses a circle to indicate each orbital and an arrow for each electron.

Important Terms

electron configuration   orbital diagram

Example _____

1. Write the orbital diagram for aluminum.

    A neutral atom of aluminum has 13 electrons. The 1s orbital is filled with 2 electrons. The 2s orbital is filled with 2 electrons. Each of the 2p orbitals will hold 2 electrons, making a total of 6 electrons in the 2p orbitals. That makes 8 electrons on the second energy level, so it is now filled. We have placed 10 electrons and have 3 left to assign to the third energy level. The 3s will be filled with 2 electrons and the last electron will be in a 3p orbital.

	1s	2s	$2p_x$	$2p_y$	$2p_z$	3s	$3p_x$	$3p_y$	$3p_z$
Aluminum	⊕	⊕	⊕	⊕	⊕	⊕	↑	○	○

2. Write the electron configuration of aluminum.

    Aluminum $1s^2 2s^2 2p^6 3s^2 3p^1$

Self-Test _____

For questions 13 to 18, fill in the blank with the correct word or words.

13. The importance of the theory of Niels Bohr is that the electrons in the atom are restricted to definite energy positions called _____ or _____.

14. The energy levels are identified by a _____.

15. An electron in the fourth energy level has _____ (more, less) energy than an electron in the third energy level.

16. For an electron to move from the second to the third energy level, it must _____ (gain, lose) energy.

17. What is the maximum number of electrons in the first energy level? _____ The third energy level? _____ The fifth energy level? _____

18. The arrangement of electrons in an atoms is called the _____.

For 19 to 23, give the symbol and atomic number, and write the electron configurations for the elements.

19. magnesium
20. neon
21. cesium

22. zinc
23. tin

3.7   Formation of Ions

If all the electrons in an atom are in the lowest possible energy levels, the atom is said to be in the ground state. Electrons in an atom can absorb energy from an exterior source and jump to a higher energy level. When this happens, the atom is said to be in an excited state. Atoms in the excited state are unstable and return to the ground state by giving off radiant energy. A positive ion is formed when one or more electrons absorb enough energy to jump completely away from the atom. Negative ions are formed when additional electrons are added to a neutral atom. Positive atoms are called cations and negative atoms are called anions.

Important Terms

ground state	excited atom	ion
cation	anion	

Self-Test _____

For questions 24 to 26, fill in the blank with the correct word or words.

24. An atom can return to the ground state by _____ (gaining, losing) energy.

25. When an atom absorbs enough energy for an electron to jump completely away from the atom, a _____ _____ is formed.

26. A negative ion, called an _____, is formed when an atom _____ electrons; a positive ion, called a _____, is formed when an atom _____ electrons.

39

3.8   The Periodic Table

Nineteenth century scientists observed that regular, repeating characteristics existed among the known chemical elements if these elements were arranged according to increasing atomic weight.   Dmitri Mendeleev was the first to publish a table of elements showing the repeating nature, or periodicity, of the chemical properties of the elements. He predicted that the gaps in his table would be filled by elements not yet discovered. Mendeleev's table was refined by arranging the elements according to atomic number rather than atomic weight. The periodic table is a valuable tool to the scientist. It is important that you become familiar with its organization and be able to make use of the information it provides.

Important Terms

    periodic table   periodicity

3.9   Periods and Groups

The horizontal rows of the periodic table are called periods. The number of the period corresponds to the outermost energy level that contains electrons for the atoms of that period. Each vertical column is called a group or chemical family. The chemical behavior of an atom is determined by the number of electrons in the outermost energy level (the valence electrons.) Every member of a chemical family has the same number of valence electrons.

The eight chemical families called the representative elements are designated by the Roman numerals IA to VIIIA. These Roman numerals identify the number of valence electrons in the atoms of each member of that chemical family. Members of the same chemical family will exhibit very similar chemical behavior.

The three rows of ten elements in the center of the table (the B group elements) are called the transition elements or transition metals. The two rows of fourteen elements at the bottom of the table are called the inner transition elements.

Some of the groups or families of elements on the periodic table are known by common names: group IA—alkali metals, group IIA—alkaline earth metals, and group VIIA—the halogens. The elements in group VIIIA, the noble gases, are unusual in their lack of chemical reactivity. This stability results from the number of valence electrons in these atoms (eight for each element except helium).

Important Terms

period   representative element   transition metal
group   inner transition element   alkali metal
family   alkaline earth metal   halogen
noble gas   valence electron

3.10 Metals, Nonmetals, Metalloids

The elements on the periodic table can be divided into two large
classes: metals and nonmetals. The metals make up by far the
majority of elements. A jagged line on the periodic table
divides the metals from the nonmetals, but there is actually no
sharp distinction between the properties of the elements
immediately on either side of this line. These elements, called
metalloids, have properties of both metals and nonmetals.

Important Terms

Metal   Nonmetal   Metalloid   Semiconductor

Self-Test

For questions 27 to 38, fill in the blank with the correct word
or words.

27. The repeating nature of the chemical properties of the
     elements is called _____.

28. The modern periodic table arranges the elements according to
     their _____.

29. A vertical column on the periodic table is called a _____
     or _____ .

30. A horizontal row on the periodic table is called a _____.

31. The number of a row indicates _____.

32. The Roman numeral above a vertical column indicates _____
     _____.

33. The electrons in the outermost energy level of an atom are
     called _____ electrons.

34. Iodine (atomic number 53) will have _____ electrons in the
     outermost energy level, which is the _____ energy level.

35. The representative elements are those in groups _____.

36. Elements in group _____, called the noble gases, are chemically _____. This results from the number of their _____ electrons.

37. Elements in the same chemical family will have similar _____.

38. An element that can be easily bent is most likely a _____, while one that is brittle is probably a _____.

Match items 39 to 48 with the correct area on the following periodic table.

39. transition metals
40. nonmetals
41. hydrogen
42. alkali metals
43. inner transition elements

44. all metals
45. halogens
46. noble gases
47. metalloids
48. alkaline earth metals

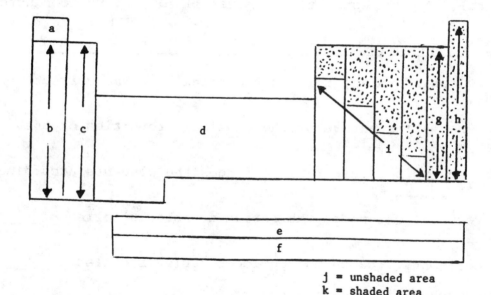

j = unshaded area
k = shaded area

3.11 The Periodic Law

The Periodic Law states that the physical and chemical properties of the elements will show a repeating nature or periodicity when the elements are arranged according to their atomic numbers. Three properties—atomic size, ionization energy, and electron affinity—all illustrate this periodicity. Understanding how these factors change on the periodic table will help you understand the chemical behavior of metals and nonmetals.

In general, the size of an atom decreases across a period and increases down a group. Metals tend to have large atomic radii, and nonmetals small radii. The ionization energy of an element increases as atomic number increases across a period, and decreases as atomic number increases down a family. Metals tend to have low ionization energies, while nonmetals have high ionization energies. The noble gases have unusually high ionization energies because of the very stable arrangement of their valence electrons. The electron affinities of the metals are low and the electron affinities of the nonmetals are high.

Important Terms

Periodic Law   ionization energy   electron affinity

Self-Test ───

To answer questions 49 to 51, use only the periodic table on the front inside cover of your text.

49. For each of the following, choose the element with the largest radius.
    (a) Se, K, Cr   (c) Ge, F, Cs   (b) Bi, N, As

50. For each of the following, choose the element with the higher ionization energy.
    (a) Li or Be   (c) Br or Kr
    (b) C or Sn   (d) Mg or S

51. Which will have a higher electron affinity, sodium or sulfur?

3.12 Elements Necessary for Life

Of the 90 naturally occurring elements, 25 are known to be essential to life. Of these 25, carbon, hydrogen, oxygen, and nitrogen are the most plentiful, making up 99.3% of all the atoms in your body. Although the remaining elements make up only 0.7% of the atoms in your body, they are critical to life. The lack of any one of them may cause disease or death.

We divide these 21 remaining elements into two groups: the macrominerals and the trace elements. The macrominerals include potassium, magnesium, sodium, calcium, phosphorus, sulfur, and chlorine. The trace elements, shown in Table 3.7, are found in very small amounts in the body. There is a specific range of concentration for each trace element that allows the body to function normally. Below that range a deficiency disease can develop. Concentrations above the range can be toxic.

Important Terms

macromineral   trace element   deficiency disease

Self-Test _____

For questions 52 to 54, fill in the blank with the correct word or words.

52.   The four elements that are most abundant in living organisms
      are _____, _____, _____, and _____.

53.   The next most abundant group of elements in the body
      contains _____ elements and together they are called the
      _____.

54.   The _____ elements exist in living organisms in minute
      amounts. These elements are made up of _____ metals and
      _____ nonmetals.

Answers to Self-Test Questions in Chapter 3

1. protons, neutrons **2.** protons **3.** atomic number **4.** mass number **5.** electron, neutron **6.** 1-, 1+, 0 **7.** (a) 1, 1, 1 (b) 6, 7, 6 (c) 8, 10, 8 (d) 14, 14, 14 (e) 35, 44, 35 **8.** F **9.** F **10.** T **11.** T **12.** 28.11 amu **13.** energy levels, energy shells **14.** number **15.** more **16.** gain **17.** 2, 18, 50 **18.** electron configuration **19.** Mg, 12, $1s^2$, $2s^2$, $2p^6$, $3s^2$ **20.** Ne, 10, $1s^2$, $2s^2$, $2p^6$ **21.** Cs, 55, $1s^2$, $2s^2$, $2p^6$, $3s^2$, $3p^6$, $4s^2$, $3d^{10}$, $4p^6$, $5s^2$, $4d^{10}$, $5p^6$, $6s^2$ **22.** Zn, 30, $1s^2$, $2s^2$, $2p^6$, $3s^2$, $3p^6$, $4s^2$, $3d^{10}$, $4p^6$, $5s^2$, $4d^{10}$ **23.** Sn, 50, $1s^2$, $2s^2$, $2p^6$, $3s^2$, $3p^6$, $4s^2$, $3d^{10}$, $4p^6$, $5s^2$, $4d^{10}$, $5p^2$ **24.** losing **25.** positive ion **26.** anion, gains, cation, loses **27.** periodicity **28.** atomic numbers **29.** group, family **30.** period **31.** the number of the outermost energy level containing electrons **32.** the number of electrons in the outermost energy level **33.** valence **34.** 7, 5th **35.** groups IA-VIIIA **36.** VIIIA, unreactive, valence **37.** chemical properties **38.** metal, nonmetal **39.** d **40.** k **41.** a **42.** b **43.** e,f **44.** j **45.** g **46.** h **47.** i **48.** c **49.** (a) K (b) Bi (c) Cs **50.** (a) Be (b) C (c) Kr (d) S **51.** sulfur **52.** carbon, hydrogen, oxygen, nitrogen **53.** 7, macrominerals **54.** trace, 8, 6

Chapter 4   COMBINATIONS OF ATOMS

In Chapter 3 we stated that the outer-shell electrons determine how atoms will interact with one another. In this chapter we examine these interactions between atoms. Atoms of elements in groups IA-VIIA will accept, share, or donate outer-shell electrons to attain eight valence electrons and become more stable. In doing so, these atoms form chemical bonds, the subject of this chapter.

4.1   The Octet Rule

The octet rule states that atoms of elements in groups IA-VIIA tend to combine with other atoms in such a way as to attain eight valence electrons.

Important Term

      octet rule

4.2   The Ionic Bond

The first type of chemical bond discussed in this chapter is the ionic bond. An ionic bond results from the force of attraction between ions that are formed when electrons are transferred from one atom to another. Ionic bonds form between atoms that strongly attract additional electrons (the nonmetals) and atoms that have weak attraction for their valence electrons (the metals). These atoms will gain or lose electrons to form a stable octet of electrons in their outer energy level. An ionic compound does not exist as separate molecules, but as groups of ions attracted to one another in an orderly arrangement called a crystal lattice. Ionic compounds are electrically neutral, so that the number of positive and negative charges in the lattice must be equal.

Important Terms

      ionic bond   ionic compound   crystal lattice

Example

The metal lithium will react with bromine to form the ionic compound lithium bromide.

Lithium needs to lose one electron to reach eight electrons in its outer shell, and bromine needs to gain one electron. A crystal of lithium bromide will contain an equal number of positive lithium ions (Li^+) and negative bromide ions (Br^-).

4.3 Lewis Electron Dot Diagrams

The electron dot diagram is an easy way to represent the number of valence electrons in atoms of the representative elements.

Important Term

electron dot diagram

Example _____

Sulfur is in group VIA and has six valence electrons. The electron dot diagram for sulfur is

$$\cdot \overset{\cdot\cdot}{\underset{\cdot}{S}} :$$

Self-Test _____

1. Write the electron dot diagrams for the following elements:
    (a) sodium
    (b) calcium
    (c) phosphorus
    (d) aluminum
    (e) bromine
    (f) argon
    (g) carbon
    (h) oxygen

2. Use Lewis electron dot diagrams to show the reaction between the following elements:
    (a) calcium and oxygen
    (b) calcium and bromine

4.4 Chemical Formulas

Chemical formulas indicate the type of atoms or ions present. The formula of a compound gives the smallest whole-number ratio of these atoms or ions that will assure electrical neutrality. For all formulas, the type of atom or ion is indicated by the symbol of the element, and the ratio of atoms or ions is shown by the use of subscripts following the symbol.

Important Term

chemical formula

Example _____

Write the chemical formula for the compound formed between aluminum and chlorine.

The formula for a compound must be electrically neutral, so the total number of positive charges must equal the total number of negative charges. Each aluminum atom needs to lose three valence electrons to become stable. Aluminum will form the Al^{3+} ion. Each chlorine atom needs to gain only one electron to reach an octet of valence electrons. Chlorine, then, will form the Cl^- ion. To maintain neutrality there must be three chloride ions for every aluminum ion. Therefore, the formula is $AlCl_3$.

Self-Test _____

For questions 3 to 7, fill in the blank with the correct word or words.

3. _____ bonds result when metals and nonmetals interact.

4. An _____ is formed by a group of positive and negative ions.

5. When forming an ionic bond, metal atoms will _____ electrons, and nonmetal atoms will _____ electrons.

6. Metals form _____ ions and nonmetals form _____ ions.

7. Chemical formulas indicate the _____ and _____ of the atoms or ions present in a compound.

8. Write the chemical formula for the compound formed between:
   (a) calcium and oxygen   (b) calcium and bromine

4.5   The Covalent Bond

Ionic bonds form between elements that have large differences in their ability to attract electrons. Such a transfer of electrons cannot take place between two elements that have similar attraction for electrons (for example, two nonmetals). To become more stable, atoms of these elements share electrons, forming covalent bonds. Covalently bonded atoms form distinct units called molecules. The atoms of some nonmetallic elements do not exist separately, but rather are found in covalently bonded pairs called diatomic molecules. In a bond diagram, the covalent bond between the atoms is represented by a dash.

Important Terms

covalent bond   covalent compound
diatomic molecule   bond diagram

4.6   Multiple Bonds

In some cases, atoms must share more than two electrons to become
stable. A single covalent bond consists of two electrons (one
pair) shared between two nuclei. A double covalent bond contains
four electrons (two pairs) shared between two nuclei, and a
triple covalent bond consists of six electrons (three pairs)
shared between two nuclei.

Important Terms

single covalent bond   double covalent bond
triple covalent bond

Example _____

Write the electron dot diagram and bond diagram for the compound
formed between hydrogen and bromine.

Both of these elements need to share one electron to become
stable. The electron dot diagram and bond diagram would be:

H :Br: H–Br

Electron dot diagram   Bond diagram

Self-Test _____

For questions 9 to 14, fill in the blank with the correct word or
words.

9.   In a covalent bond, electrons are _____ between two atoms.

10. _____ bonds tend to form between a metal and a nonmetal,
     while _____ bonds form between two nonmetals.

11. Covalent compounds are composed of distinct units called
     _____.

12. Six electrons shared between two atoms is called a _____
     covalent bond, and is represented by _____ drawn between
     the symbols of the elements.

13. Two electrons shared between two atoms is a _____ covalent bond and is represented by _____ drawn between the symbols of the elements.

14. Four electrons shared between two atoms is a _____ covalent bond and is represented by _____ drawn between the symbols of the elements.

15. Write the formula, electron dot diagram, and bond diagram for the covalent compounds formed between the following elements:
    (a) carbon and iodine   (c) carbon and oxygen
    (b) nitrogen and bromine   (d) iodine and iodine

4.7   Electronegativity

Electrons in a covalent bond are not necessarily shared equally between the two nuclei. The tendency of an atom to attract shared electrons is called the electronegativity of that element. The more electronegative an element, the greater its attraction for shared electrons. On the periodic table, electronegativity increases across a period and decreases down a group.

Identical atoms will share electrons equally and will form nonpolar covalent bonds. Elements whose electronegativities differ will form polar covalent bonds, with the positive end of the bond near the less electronegative atom and the negative end of the bond near the more electronegative atom. If the electronegativity difference between the two atoms is great enough, the electrons will not be shared, but will be transferred—resulting in the formation of an ionic bond. There is no distinct separation between polar covalent bonds and ionic bonds. But for our purposes it is convenient to make the following generalization: ionic bonds form between metals and nonmetals, and covalent bonds form between nonmetals.

Important Terms

> electronegativity   polar covalent bond
> nonpolar covalent bond

Self-Test _____

For questions 16 to 19, indicate (a) which element is the more electronegative, and (b) the type of bond (ionic, nonpolar covalent, polar covalent) that will form between elements.

16. chlorine and magnesium   18. iodine and iodine
17. phosphorus and chlorine   19. bromine and potassium

4.8 Polar and Nonpolar Molecules

A molecule is nonpolar when the centers of positive and negative charge coincide, and is polar when these centers do not coincide. Whether a molecule is polar or nonpolar is determined both by the type of bonds present in the molecule (polar or nonpolar) and the arrangement of these bonds (the shape of the molecule). The nonmetallic elements in groups IVA to VIIA, when singly bonded to other elements, take on characteristic shapes. These shapes are illustrated in Figure 4.11 in the text.

Important terms

nonpolar molecule   polar molecule

Self-Test _____

For questions 20 to 24, indicate (a) the type of bonding (polar covalent, nonpolar covalent, or ionic) found in the compound, and (b) if the bonding is covalent, indicate whether the molecule is polar or nonpolar.

20. CCl_4 21. NBr_3 22. LiBr   23. $CHBr_3$ 24. H_2

4.9 Hydrogen Bonding

A hydrogen bond is an attraction that forms between a hydrogen atom bonded to a highly electronegative atom on one molecule and a highly electronegative atom on another molecule (or on another region of the same molecule). A hydrogen bond is a fairly weak attraction, equal to only about one-tenth that of an ordinary covalent bond. But hydrogen bonds are critical in influencing the behavior of many biologically important molecules.

Important Terms

hydrogen bond   intermolecular   intramolecular

Self-Test _____

25. In which three of the following might a hydrogen bond form?
   (a) When a hydrogen atom is bonded to an iodine atom.
   (b) When a hydrogen atom is bonded to an oxygen atom.
   (c) When a hydrogen atom is bonded to a nitrogen atom.
   (d) Between a hydrogen atom on one molecule and a hydrogen atom on another molecule.
   (e) Between an oxygen atom on one molecule and a hydrogen atom on another molecule.

4.10 Polyatomic Ions

A polyatomic ion is a charged group of atoms that is held together as a stable unit by covalent bonds. This unit remains intact through most chemical reactions. Polyatomic ions will form ionic bonds with metals.

Important Term

polyatomic ion

4.11 Oxidation Number

Oxidation numbers are a helpful way of keeping track of the transfer of electrons in chemical reactions. They also help in writing chemical formulas. The following rules are used when assigning oxidation numbers to elements:

Rule 1   Elements have an oxidation number of zero when they are not combined with atoms of a different element.

Rule 2   In a covalent bond, the oxidation numbers are assigned as if the electrons were transferred to the more electronegative atom.

Rule 3   An ion containing only one atom has a oxidation number equal to the charge on the ion.

Rule 4   The sum of the oxidation numbers of the atoms in a chemical compound is zero. The sum of the oxidation numbers of the atoms in a polyatomic ion equal the charge on that ion.

Important Terms

oxidation number   oxidation state   binary compound

Example

What is the oxidation number of each element in the following:

(a) CaO   This is an ionic compound in which the calcium forms a Ca^{2+} ion and the oxygen forms a O^{2-} ion. Using rule 3, the oxidation number of calcium is 2+ and the oxidation number of oxygen is 2-.

(b) HBr   This is a covalent compound in which the bromine is the more electronegative atom. Using rule 2, the oxidation number of bromine is 1- and the oxidation number of hydrogen is 1+.

(c) Na_2CO_3 This compound contains more than two elements, so
we determine the answer by assigning oxidation
numbers to the elements we know, and then
determining the oxidation number of the remaining
element.

$$Na: (1+) \times 2 \text{ atoms} = 2+$$
$$C: (X) \times 1 \text{ atom} = X$$
$$O: (2-) \times 3 \text{ atoms} = 6-$$

For the sum of the oxidation numbers to be zero, X must
equal 4+; therefore, the oxidation number of the carbon
atom in this compound is 4+.

Self-Test _____

26. Determine the oxidation number of each element in the
following compounds and ions:
(a) NaI   (c) K_3PO_4 (e) HSO_3^-
(b) N_2 (d) H_2O (f) $LiClO_3$

4.12 Naming Chemical Compounds

Binary ionic compounds are named by placing the name of the
positive ion first and the name of the negative ion second. The
positive ion carries the name of the parent element. Roman
numerals are placed after the name of a metal to indicate the
oxidation number if the metal forms more than one positive ion.
The negative ion uses the name of the parent element with an -ide
suffix. Ionic compounds containing a metal and a polyatomic ion
are named by placing the name of the metal first and the name of
the polyatomic ion second.

Binary covalent compounds are named by writing the name of the
element with the lower electronegativity first. The second
element is named by adding the suffix -ide to the name of the
parent element. Prefixes are used to indicate the number of
atoms of each element present in the molecule.

Example _____

Write the name of the following compounds:

(a) $ZnCl_2$ This is an ionic compound formed between the metal
zinc and the nonmetal chlorine. The name of the
metal ion is placed first with the nonmetal ion
name ending with the suffix -ide: zinc chloride.

52

(b) CuS We note in Table 4.3 that copper has two oxidation numbers: 1+ and 2+. To determine which copper ion this is, we use the fact that in binary compounds sulfur has an oxidation number of 2-. Therefore, this compound must contain the copper (II) ion, Cu^{2+}. The name of this compound is copper(II) sulfide.

(c) CS_2 This is a binary covalent compound. Sulfur is the more electronegative atom, so carbon is the element to be named first. There are two sulfur atoms, so the prefix "di" is used. The name is carbon disulfide.

Self-Test ───

For questions 27 to 46, name each compound.

27. CuCl

28. SO_3

29. K_2S

30. $FeBr_2$

31. SiO_2

32. NaI

33. N_2O_4

34. $NaHSO_3$

35. $Mg_3(PO_4)_2$

36. KCN

37. HBr

38. Al_2S_3

39. CuO

40. H_2S

41. $CaCl_2$

42. SF_6

43. Li_2O

44. $(NH_4)_2SO_4$

45. $LiClO_3$

46. Na_2O_2

4.13 Writing Chemical Formulas

Chemical formulas of compounds are written to show the lowest whole-number ratio of atoms that will assure electrical neutrality. The symbol of the atom with the more positive oxidation number is written first. If a chemical formula contains more than one polyatomic ion, the symbol for the polyatomic ion is placed in parentheses, followed by a subscript indicating the number of polyatomic ions.

Example ───

Write the chemical formula for each of the following:

(a) **Barium chloride:** From Table 4.2, we see that barium forms a Ba^{2+} ion and chlorine forms a Cl^- ion. To maintain neutrality, there must be two chlorine atoms for every barium atom. The formula is $BaCl_2$.

(b) **Aluminum oxide:** Again from Table 4.2, aluminum will form an Al^{3+} ion and oxygen an O^{2-} ion. To maintain neutrality, we need two aluminum ions (a total charge of 6+) and three oxide ions (a total charge of 6-). The formula is Al_2O_3. Using the crisscross method to determine the formula, we have

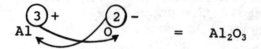

$$= \quad Al_2O_3$$

(c) **Calcium nitrate:** This is an ionic compound formed between the calcium ion (Ca^{2+}) and the polyatomic nitrate ion (NO_3^-). Using the crisscross method, we have

$$= \quad Ca(NO_3)_2$$

Self-Test _____

For questions 47 to 66, write the chemical formula for each compound.

47. silver sulfide
48. nitrogen dioxide
49. copper(I) acetate
50. aluminum sulfate
51. iron(III) oxide
52. magnesium nitride
53. dinitrogen pentoxide
54. potassium dichromate
55. nitrogen trichloride
56. barium sulfide

57. sodium fluoride
58. carbon tetrachloride
59. ammonium chloride
60. iron(II) sulfite
61. zinc bromide
62. aluminum oxide
63. hydrogen selenide
64. magnesium hydroxide
65. chlorine trifluoride
66. copper(I) oxide

1. (a) Na (b) Ca (c) P (d) Al (e) Br (f) Ar (g) C (h) O

2. (a) Ca + O ——————▶ Ca^{2+} O^{2-}

   (b) Br + Ca + Br ——————▶ Br^- Ca^{2+} Br^-

3. ionic 4. ionic compound (crystal lattice) 5. lose, gain
6. positive, negative 7. type, ratio 8. (a) CaO (b) $CaBr_2$
9. shared 10. ionic, covalent 11. molecules 12. triple, three
dashes 13. single, dash 14. double, two dashes
15.

(a) CI_4

(b) NBr_3

(c) CO_2

(d) I_2

16. (a) chlorine (b) ionic 17. (a) chlorine (b) polar covalent
18. (a) the same (b) nonpolar covalent 19. (a) bromine (b) ionic
20. (a) polar (b) nonpolar 21. (a) polar (b) polar
22. (a) ionic 23. (a) polar (b) polar 24. (a) nonpolar
(b) nonpolar 25. b,c,e 26. (a) Na 1+; I 1- (b) N 0 (c) K 1+;
P 5+; O 2- (d) H 1+; O 2- (e) H 1+; S 4+; O 2- (f) Li 1+; Cl 5+;
O 2- 27. copper(I) chloride 28. sulfur trioxide 29. potassium
sulfide 30. iron(II) bromide 31. silicon dioxide 32. sodium
iodide 33. dinitrogen tetroxide 34. sodium hydrogen sulfite
35. magnesium phosphate 36. potassium cyanide 37. hydrogen
bromide 38. aluminum sulfide 39. copper(II) oxide 40. hydrogen
sulfide 41. calcium chloride 42. sulfur hexafluoride
43. lithium oxide 44. ammonium sulfate 45. lithium chlorate
46. sodium peroxide 47. Ag_2S 48. NO_2 49. $CuC_2H_3O_2$
50. $Al_2(SO_4)_3$ 51. Fe_2O_3 52. Mg_3N_2 53. N_2O_5 54. $K_2Cr_2O_7$ 55. NCl_3
56. BaS 57. NaF 58. CCl_4 59. NH_4Cl 60. $FeSO_3$ 61. $ZnBr_2$
62. Al_2O_3 63. H_2Se 64. $Mg(OH)_2$ 65. ClF_3 66. Cu_2O

Chapter 5   CHEMICAL EQUATIONS AND THE MOLE

In this chapter we discuss writing and balancing equations that represent chemical reactions. A second important concept introduced is the mole, the reacting unit used in chemistry. By the time you have finished this chapter, it is important that you feel comfortable with the definition of the mole and calculations using the mole.

5.1   Writing Chemical Equations

Chemical equations are a shorthand method of describing a chemical reaction. The substances that you begin with are called the reactants, and appear to the left of the arrow in the chemical equation. The substances that are formed in a chemical reaction are called the products, and appear to the right of the arrow in the chemical equation.

When chemical reactions occur, atoms are neither created nor destroyed—only rearranged. Therefore, the number of atoms of each element on the reactant side of the equation must equal the number of atoms of that element on the product side of the equation. An equation written this way is said to be balanced. In a chemical reaction, the total mass of the reactants will equal the total mass of the products (the Law of Conservation of Mass).

Important Terms

        reactant   product
        Law of Conservation of Mass   balanced equation

5.2   Balancing Chemical Equations

Study carefully the five steps described in your text for balancing chemical equations. You might want to refer to them when reading the following examples and then try each example on your own. Confidence in balancing chemical equations comes only with lots and lots of practice.

Example _____

1.   Magnesium reacts with oxygen to produce magnesium oxide.

Step 1: $Mg + O_2 \longrightarrow MgO$

Steps 2, 3, 4: Looking at the above equation we see that the magnesium atoms balance, but not the oxygen atoms. In order to have the same number of oxygen atoms on both sides of the equation we need to produce two magnesium oxides. (Remember that once you have written the correct chemical formulas for the reactants and products, you only adjust the coefficients to balance the equation.)

$$Mg + O_2 \longrightarrow 2MgO$$

Now the oxygen atoms balance, but not the magnesium atoms. Since there are two magnesium atoms on the product side, there must be two magnesium atoms on the reactant side.

$$2Mg + O_2 \longrightarrow 2MgO$$

Step 5: We have two magnesium atoms on each side and two oxygen atoms on each side, so our equation is balanced.

2. Sodium hydroxide will react with ferric chloride to produce ferric hydroxide and sodium chloride.

Step 1: It is critical that you spend time to get each chemical formula correct because the equation cannot be balanced properly if a formula is wrong.

$$NaOH + FeCl_3 \longrightarrow Fe(OH)_3 + NaCl$$

If you have trouble writing these formulas, reread Sections 4.4 and 4.13 of the text.

Steps 2, 3, 4: Begin by giving $Fe(OH)_3$ a coefficient of 1. The iron atoms balance. There are three hydroxides (OH) on the product side, so we need three hydroxide ions on the reactant side.

$$3NaOH + FeCl_3 \longrightarrow Fe(OH)_3 + NaCl$$

Now balance the sodium atoms by putting a coefficient of three before the NaCl on the product side. Check the chlorine atoms. Do they balance?

$$3NaOH + FeCl_3 \longrightarrow Fe(OH)_3 + 3NaCl$$

Step 5: Check this equation yourself. Is it balanced?

3. Balance the following equation:

$$4P + 5O_2 \longrightarrow 2P_2O_5$$

<u>Step 1:</u> Completed

<u>Steps 2, 3, 4:</u> Balance the phosphorus.

$$2P + O_2 \longrightarrow P_2O_5$$

Balancing the oxygen takes a little thinking. There are five oxygen atoms on the product side, so we need five oxygen atoms on the reactant side. Each oxygen molecule contains two oxygen atoms, we need 2 1/2 (or 5/2) oxygen molecules.

$$2P + \frac{5}{2} O_2 \longrightarrow P_2O_5$$

This equation is balanced, but we would prefer to have whole number coefficients in our equation. To eliminate the fraction, we can multiply the entire equation by 2 and the equation will remain balanced.

$$4P + 5O_2 \longrightarrow 2P_2O_5$$

<u>Step 5:</u> Is the equation balanced? Check for yourself.

Self-Test _____

For questions 1 to 5, write a sentence saying in words what is indicated by the chemical equations.

1. $H_2 + S \longrightarrow H_2S$ *hydrogen sulfur*
2. $Ca + 2H_2O \longrightarrow Ca(OH)_2 + H_2$ *Calcium hydroxid + hydrogen*
3. $Cu + 2AgNO_3 \longrightarrow 2Ag + Cu(NO_3)_2$ *2 Silver 2 copper nitrate*
4. $2H_2O + C \longrightarrow CO_2 + 2H_2$ *Carbon dioxide 2 hydrogen*
5. $F_2 + 2NaCl \longrightarrow 2NaF + Cl_2$ *2 sodium flour + 1 chlorine*

For questions 6 to 20, balance the equations.

6. $2HCl + Fe \longrightarrow FeCl_2 + H_2$
7. $N_2 + 2O_2 \longrightarrow 2NO_2$
8. $2H_3PO_4 + 3Ca(OH)_2 \longrightarrow Ca_3(PO_4)_2 + 6H_2O$
9. $3Pb + Al_2O_3 \longrightarrow 2Al + 3PbO$
10. $4Al + 3O_2 \longrightarrow 2Al_2O_3$
11. $CH_4 + 2O_2 \longrightarrow CO_2 + 2H_2O$
12. $3Ni + 2As \longrightarrow Ni_3As_2$
13. $2Al + N_2 \longrightarrow 2AlN$
14. $2C_2H_2 + 5O_2 \longrightarrow 4CO_2 + 2H_2O$
15. $3Mg + Fe_2O_3 \longrightarrow 3MgO + 2Fe$

16. Potassium reacts with oxygen to form potassium oxide. $4K + O_2 \rightarrow K_2O$
17. Tin reacts with bromine to form tin(Il) bromide. $Sn + Br_2 \rightarrow SnBr_2$
18. Magnesium + nitrogen ⟶ magnesium nitride $3Mg + N_2 \rightarrow Mg_3N_2$
19. Hydrogen sulfide + oxygen ⟶ sulfur dioxide + water $2H_2S + 3O_2 \rightarrow 2SO_2 + 2H_2O$
20. Calcium hydroxide + sodium carbonate ⟶ sodium hydroxide + calcium carbonate

$$Ca(OH)_2 + Na_2CO_3 \rightarrow 2NaOH + CaCO_3$$

5.3 Using Oxidation Numbers to Balance Oxidation-Reduction Equations (optional section)

Chemical reactions that involve the transfer of electrons are called oxidation-reduction or redox reactions. Oxidation is the loss of electrons by an atom and reduction is the addition of these electrons to another atom. In redox reactions, the element that gains electrons is the oxidizing agent and the element that loses electrons is the reducing agent.

In a chemical reaction the number of electrons lost must equal the number of electrons gained. This fact allows us to use oxidation numbers to help balance the equations for redox reactions.

Important Terms

oxidation   reduction   redox reaction
oxidizing agent   reducing agent

Example _____

Balance the following equation using oxidation numbers and identify the oxidizing and reducing agents.

$$KMnO_4 + FeSO_4 + H_2SO_4 \longrightarrow MnSO_4 + Fe_2(SO_4)_3 + K_2SO_4 + H_2O$$

To identify the elements that change oxidation numbers (those elements that gain or lose electrons in this reaction), we must assign oxidation numbers to all of the elements. If you are uncertain how to do this, review Section 4.11 in your text.

$KMnO_4$	K	1+		$MnSO_4$	Mn	2+
	Mn	7+			SO_4	2-
	O	2-				
				$Fe_2(SO_4)_3$	Fe	3+
$FeSO_4$	Fe	2+			SO_4	2-
	SO_4	2-				
				K_2SO_4	K	1+
					SO_4	2-

59

H_2SO_4	H	1+	H_2O	H	1+
	SO_4	2−		O	2−

(Because the polyatomic sulfate ion remains unchanged in this reaction, we can treat it as a single unit with an oxidation number of 2−)

We can see that two elements change oxidation numbers: Mn and Fe. Mn gains electrons, so it is the oxidizing agent. Fe loses electrons, so it is the reducing agent. Isolating these elements from the equation for a moment, we have:

$$KMnO_4 \ + \ 2FeSO_4 \ \longrightarrow \ MnSO_4 \ + \ Fe_2(SO_4)_3$$

$$\underset{7+}{} \qquad \underset{2+}{} \qquad\qquad \underset{2+}{} \qquad \underset{3+}{}$$

5 e gained
2 × (1e lost per Fe) = 2e lost

To have the number of electrons lost equal to the number of electrons gained, we must determine the lowest common denominator between the number lost and the number gained. This would be 10. To have 10 electrons gained we must have two Mn on each side of the equation. To have ten electrons lost we need ten Fe atoms on each side of the equation.

$$2KMnO_4 + 10FeSO_4 + H_2SO_4 \longrightarrow 2MnSO_4 + 5Fe_2(SO_4)_3 + K_2SO_4 + H_2O$$

Now we must balance the other atoms. Check the potassium ions. They are balanced. Next check the sulfate ions. On the product side there are 18 sulfate ions. On the reactant side, there are 10 sulfates in 10$FeSO_4$ and this coefficient must not be changed. To supply the 8 other sulfate ions, we must have 8H_2SO_4.

$$2KMnO_4 + 10FeSO_4 + 8H_2SO_4 \longrightarrow 2MnSO_4 + 5Fe_2(SO_4)_3 + K_2SO_4 + H_2O$$

Now we need to balance the hydrogen atoms. We have 16 hydrogen atoms on the reactant side of the equation. To have 16 hydrogen atoms on the product side we must have 8H_2O.

$$2KMnO_4 + 10FeSO_4 + 8H_2SO_4 \longrightarrow 2MnSO_4 + 5Fe_2(SO_4)_3 + K_2SO_4 + 8H_2O$$

The last element to balance is the oxygen; the oxygen atoms balance—8 atoms of oxygen on each side of the equation. The equation is now completely balanced.

Self-Test

For questions 21 to 26, balance the equations using oxidation numbers.

21. $FeCl_3$ + $SnCl_2$ $\longrightarrow$ $FeCl_2$ + $SnCl_4$

22. H_2SO_3 + HNO_3 $\longrightarrow$ H_2SO_4 + NO + H_2O

23. $KMnO_4$ + KNO_2 + H_2SO_4 $\longrightarrow$ $MnSO_4$ + H_2O + KNO_3 + K_2SO_4

24. HCl + $KMnO_4$ $\longrightarrow$ H_2O + KCl + $MnCl_2$ + Cl_2

25. MnO_2 + HCl $\longrightarrow$ H_2O + $MnCl_2$ + Cl_2

26. H_2S + O_2 $\longrightarrow$ SO_2 + H_2O

5.4 The Mole

This section introduces a key concept—the reacting unit in chemistry—the mole. The mole allows scientists to control the exact amount of each element or compound used in the laboratory. Since atoms are so small, it is impossible to measure out one thousand or even one million atoms of an element from a reagent bottle. Therefore, an arbitrary unit was established that allows scientists to measure an exact number of atoms. This unit, the mole, contains 6.02×10^{23} units (the units themselves depend on the specific nature of the substance). One mole of potassium ions contains 6×10^{23} ions, one mole of water contains 6×10^{23} water molecules, one mole of magnesium contains 6×10^{23} atoms, and one mole of donuts contains 6×10^{23} donuts.

The weight of a mole depends upon the nature of the substance involved. For example, one mole or 6×10^{23} peas would weigh significantly less than one mole or 6×10^{23} watermelons. The weight of one mole of atoms of an element is defined as the atomic weight of that element in grams. Therefore, one mole of magnesium will weigh 24 grams, and one mole of helium will weigh four grams.

Note: In problems involving calculations with the mole, we will use atomic weights to three significant figures.

Important Terms

mole (mol)   Avogadro's number

5.5 Formula Weight

The formula weight of a substance equals the sum of the atomic weights of all the atoms in the formula of that substance. The formula weight of chlorine (Cl_2) is 35.5 + 35.5 = 71.0 amu. The formula weight of hydrochloric acid (HCl) is 1.0 + 35.5 = 36.5 amu. The formula weight of calcium hydroxide [$Ca(OH)_2$] is calculated by adding the atomic weight of calcium to two times the sum of the atomic weight of hydrogen plus the atomic weight of oxygen: 40.1 + 2(1.0 + 16.0) = 40.1 + 34.0 = 74.1 amu.

One mole of any substance will have a mass equal to the formula weight of that substance in grams. One mole of chlorine weighs 71.0 grams, one mole of hydrochloric acid weighs 36.5 grams, and one mole of calcium hydroxide weighs 74.1 grams.

Important Term

    formula weight

Self-Test

For questions 27 to 36, choose the correct answer. (Keep your periodic table handy to help you with these problems.)

27. One mole is equal to the number of atoms in
    (a) 12.0000 grams of the carbon-12 isotope.
    (b) 6×10^{23} grams of the carbon-12 isotope
    (c) 6×10^{23} grams of any element
    (d) 12.000 grams of any element

28. Avogadro's number refers to
    (a) the weight of one mole of any substance
    (b) the number of atoms in 1 gram of any element
    (c) the number of atoms in a mole
    (d) 6×10^{23} grams of any substance

29. The weight of one mole of any substance equals
    (a) Avogadro's number
    (b) the formula weight in grams
    (c) the atomic weight in grams
    (d) 6×10^{23} grams

30. The weight of one mole of potassium atoms is.
    (a) 78.2 g   (c) 31.0 g
    (b) 39.1 g   (d) 12.0 g

31. The weight of 2.80 moles of beryllium atoms is
    (a) 9.01 g   (c) 18.0 g
    (b) 27.3 g   (d) 25.2 g $280 \text{ moles} \times \dfrac{9.01 g}{1 \text{mole}} = 25.2g$

32. The weight of 0.360 mole of argon atoms is
    (a) 14.4 g   (c) 40.0 g
    (b) 144 g   (d) 111 g $0.360 \text{mole} \times \dfrac{39.95g}{1\text{mole}} = 14.38 g$

33. The weight of 3×10^{22} atoms of helium is
    (a) 2 g   (c) 0.2 g
    (b) 0.4 g   (d) 0.02 g $3 \times 10^{22} \text{atoms} \times \dfrac{1 \text{mol}}{6.02 \times 10^{23} \text{atoms}} \times \dfrac{4.00g}{1\text{mole}} = \dfrac{1.2 \times 10^{23}}{6.02 \times 10^{23}}$

34. The formula weight of $BaCl_2$ is
    (a) 20.8 amu   (c) 81.8 amu
    (b) 172 amu   (d) 208 amu 1.99×10^{44}

$137 + 35 + 35$ 62

207.

35. The formula weight of $Al_2(CO_3)_3$ is
    (a) 87.0 amu   (c) 234 amu
    (b) 114 amu   (d) 207 amu

36. The weight of one mole of $Ca_3(PO_4)_2$ is
    (a) 310 g   (c) 135 g
    (b) 230   (d) 184 g $120 + 60 + 128$

37. What is the formula weight of each of the following?
    (a) MnO_2 $55 + 32 = 87g$ (c) CO_2 $12 + 32 = 44 AMU$ (e) $KClO_3$ $39 + 35+$ (g) H_2 $= 2.00 amu$
    (b) $Al(NO_3)_3$ AMU (d) $Be_3(PO_4)_2$ (f) H_2SO_4 $12 amu$ (h) $NaBr$ $23 + 80$
    $27 + 42 + 144 =$ $27 + 62 + 128$ $48=$ $103.00 amu$
    $213 AMU$ $217 amu$ $2 + 32 + 64 = 98 amu$

5.6   Using the Mole in Problem Solving

Doing calculations using the mole can be simplified if you use the unit factor method discussed in Chapter 1. If you have not done so recently, reread Section 1.10. The following steps are helpful to follow when solving problems involving the mole.

Step 1:   Establish a relationship between the given quantity in the problem and the unknown quantity. (In some problems this may require more than one equation.)

Step 2:   Write unit factors from the equation (or equations) in Step 1.

Step 3:   Determine what the question is asking for so that you can choose the correct unit factor (or factors) to use.

Step 4:   Solve the problem, making certain to cancel units so that the answer is in the correct units.

Example ——————————————————————————————

1.   How much does 0.0240 mole of calcium hydroxide weigh?

   Step 1:   1 mol $Ca(OH)_2$ =   74.1 g

   Step 2: $\dfrac{1 \text{ mol } Ca(OH)_2}{74.1 \text{ g}}$ and $\dfrac{74.1 \text{ g}}{1 \text{ mol } Ca(OH)2}$

   Step 3:   0.0240 mol $Ca(OH)_2$ =   ? grams

   Step 4:   0.0240 mol $Ca(OH)_2$ $\times \dfrac{74.1 \text{ g}}{1 \text{ mol } Ca(OH)_2}$ = 1.78 g

2.   How many moles are there in 355 milligrams of chlorine gas?

Step 1: 1 mol Cl_2 = 71.0 g and 1 g = 1000 mg (Remember: chlorine exists as the diatomic molecule Cl_2)

Step 2: $\dfrac{1 \text{ mol } Cl_2}{71.0 \text{ g}}$ and $\dfrac{71.0 \text{ g}}{1 \text{ mol } Cl_2}$ $\dfrac{1 \text{ g}}{1000 \text{ mg}}$ and $\dfrac{1000 \text{ mg}}{1 \text{ g}}$

Step 3: 355 mg = ? mol Cl_2

Step 4: 355 mg $\times \dfrac{1 \text{ g}}{1000 \text{ mg}} \times \dfrac{1 \text{ mol } Cl_2}{71.0 \text{ g}} = \dfrac{355 \text{ mol } Cl_2}{1000 \times 71.0}$

$$= 0.00500 \text{ mol } Cl_2$$

Self-Test _____

For questions 38 and 39, you will want to refer back to the formula weights that you determined in question 37.

38. How many grams are there in each of the following?
  (a) 3.00 moles of MnO_2 3.00 mole $\times \dfrac{87 g}{1 mol\ MnO_2} = 261 g$ 56 + 32
  (b) 2.00×10^{21} atoms of $Al(NO_3)_3$
  (c) 0.250 mole of CO_2
  (d) 4.50 moles of $Be_3(PO_4)_2$

39. How many moles are there in each of the following?
  (a) 2.40×10^{24} formula units of $KClO_3$
  (b) 0.482 grams of H_2SO_4
  (c) 1.00 kg of H_2
  (d) 4.60 mg of sodium bromide

5.7   Calculations Using Balanced Equations

A balanced equation tells us many things about a reaction. It tells us the number of atoms, the number of molecules of a covalent compound, the number of formula units of an ionic compound, the number of moles, even the number of grams of each substance involved in the reaction. Look at the following balanced equation:

$$Zn + 2HCl \longrightarrow ZnCl_2 + H_2$$

This equation can be read in the following ways:

1. One atom of zinc will react with two molecules of hydrochloric acid to form one formula unit of zinc chloride and one molecule of hydrogen.

2.   One mole of zinc will react with two moles of hydrochloric
     acid to form one mole of zinc chloride and one mole of
     hydrogen.

3.   65.4 grams of zinc will react with 73.0 grams of
     hydrochloric acid to form 136.4 grams of zinc chloride and
     2.0 grams of hydrogen. (Do you see that statement 3 follows
     directly from statement 2?)

All this information allows us to perform calculations to
determine unknown quantities in a reaction.

Example

1.   Assume you have 13.0 grams of zinc. (a) If you react all of
     it with hydrochloric acid, how many grams of zinc chloride
     will you produce? (b) What is the minimum number of grams
     of hydrochloric acid that must be present to react with all
     of the zinc?

     (a) Using the balanced equation for the reaction,

$$Zn \ + \ 2HCl \ \longrightarrow \ ZnCl_2 \ + \ H_2$$

     we need to write the unit factors necessary to determine how
     many grams of zinc chloride will be produced if 13.0 grams
     of zinc react. We need unit factors for the following
     conversions:

$$\text{grams Zn} \xrightarrow{\text{(1)}} \text{moles Zn} \xrightarrow{\text{(2)}} \text{moles ZnCl}_2 \xrightarrow{\text{(3)}} \text{grams ZnCl}_2$$

     (1) The first unit factor we need involves the conversion
          of grams of Zn to moles of Zn.

$$\frac{1 \text{ mol Zn}}{65.4 \text{ g Zn}} \quad \text{and} \quad \frac{65.4 \text{ g Zn}}{1 \text{ mol Zn}}$$

     (2) For the second conversion we need to look at the ratio
          of coefficients in our balanced equation.

$$\frac{1 \text{ mol Zn}}{1 \text{ mole ZnCl}_2} \quad \text{and} \quad \frac{1 \text{ mol ZnCl}_2}{1 \text{ mol Zn}}$$

     (3) For the third conversion, we need a conversion factor
          showing the ratio of moles of zinc chloride to grams of
          zinc chloride.

$$\frac{1 \text{ mol ZnCl}_2}{136.4 \text{ g ZnCl}_2} \quad \text{and} \quad \frac{136.4 \text{ g ZnCl}_2}{1 \text{ mol ZnCl}_2}$$

To solve the problem in one step, we must choose the unit factors from (1), (2), and (3) in such a way that all the units cancel except grams of $ZnCl_2$.

$$13.0 \text{ g Zn} \times \frac{1 \text{ mol Zn}}{65.4 \text{ g Zn}} \times \frac{1 \text{ mol ZnCl}_2}{1 \text{ mol Zn}} \times \frac{136.4 \text{ g ZnCl}_2}{1 \text{ mol ZnCl}_2}$$

$$= \; 27.1 \text{ g ZnCl}_2$$

(b) From the balanced equation, we see that for every mole of zinc reacted, 2 moles of hydrochloric acid will react. if we know how many moles of zinc we have, we can calculate the required moles of HCl and grams of HCl.

$$13.0 \text{ g Zn} \times \frac{1 \text{ mol Zn}}{65.4 \text{ g Zn}} \times \frac{2 \text{ mol HCl}}{1 \text{ mol Zn}} \times \frac{73.0 \text{ g HCl}}{1 \text{ mol HCl}}$$

$$= 14.5 \text{ g HCl}$$

2.   How many grams of sulfur dioxide will be produced when 500 g of hydrogen sulfide is reacted with 700 g of oxygen? (The other product is water.)

Often in the industrial production of chemicals, one of the reactants will be added in excess. The reactant that is completely reacted (and used up) before any other is called the limiting reactant. We will need to determine the limiting reactant to solve this problem.

(1)   First we need to write a balanced equation for this reaction.

$$2H_2S + 3O_2 \longrightarrow 2SO_2 + 2H_2O$$

(2)   Next, we need to determine the number of moles of each reactant.

$$500 \text{ g} \times \frac{1 \text{ mol H}_2\text{S}}{34.1 \text{ g}} \; = \; 15.6 \text{ mol H}_2\text{S}$$

$$700 \text{ g} \times \frac{1 \text{ mol O}_2}{32.0 \text{ g}} \; = \; 21.8 \text{ mol O}_2$$

(3)   From the balanced equation, we see that three moles of hydrogen sulfide will react with two moles of oxygen, so for 15.6 mol of H_2S, we need

$$15.6 \text{ mol H}_2\text{S} \times \frac{3 \text{ mol O}_2}{2 \text{ mol H}_2\text{S}} = 23.4 \text{ mol O}_2$$

Thus, we see that we don't have enough oxygen to react with all of the hydrogen sulfide. So oxygen will be the limiting reactant. Since the reaction will stop when all of the oxygen is used up, we use the number of moles of the limiting reagent, oxygen, to calculate the number of grams of sulfur dioxide that will be produced.

$$21.8 \text{ mol O}_2 \times \frac{2 \text{ mol SO}_2}{3 \text{ mol O}_2} \times \frac{64.1 \text{ g}}{1 \text{ mol SO}_2} = 932 \text{ g}$$

Self-Test

For questions 40 to 42, write the equations as sentences involving (a) atoms, molecules, and formula units; (b) moles; and (c) grams.

40. $N_2 + 3H_2 \longrightarrow 2NH_3$
41. $FeCl_3 + 3NaOH \longrightarrow Fe(OH)_3 + 3NaCl$
42. $F_2 + 2NaCl \longrightarrow 2NaF + Cl_2$

43. How many grams of water are formed when 51.0 grams of ammonia are burned in oxygen? (The other product is nitrogen gas.)

44. Chlorine is a very important industrial chemical. Many tons of chlorine are produced each year by the breakdown of sodium chloride. The other product of the reaction is sodium. How many kilograms of sodium chloride are required to produce 2.00 tons of chlorine?

45. Silver reacts with nitric acid in the following manner:

$$Ag + 2HNO_3 \longrightarrow NO_2 + AgNO_3 + H_2O$$

   (a) How many grams of HNO_3 are required to react with 27.0 grams of silver?
   (b) How many grams of NO_2 will be formed?
   (c) How many grams of $AgNO_3$ will be formed?

46. How many tons of ammonia will be produced when 1.00 ton of nitrogen gas is reacted with 0.500 ton of hydrogen gas. (Hint: you must determine which of these reactants is the limiting reactant.)

Answers to Self-Test Questions in Chapter 5

1. One hydrogen molecule reacts with one sulfur atom to produce one molecule of hydrogen sulfide. **2.** One calcium atom reacts with two water molecules to form one calcium hydroxide ion group and one molecule of hydrogen. **3.** One copper atom reacts with two silver nitrate ion groups to form two silver atoms and one copper nitrate ion group. **4.** Two water molecules and one carbon atom react to form one molecule of carbon dioxide and two molecules of hydrogen. **5.** One molecule of fluorine reacts with two sodium chloride ion groups to form two sodium fluoride ion groups and one chlorine molecule. **6.** $2HCl + Fe \rightarrow FeCl_2 + H_2$

7. $N_2 + 2O_2 \rightarrow 2NO_2$ **8.** $2H_3PO_4 + 3Ca(OH)_2 \rightarrow Ca_3(PO_4)_2 + 6H_2O$

9. $3Pb + Al_2O_3 \rightarrow 2Al + 3PbO$ **10.** $4Al + 3O_2 \rightarrow 2Al_2O_3$

11. $CH_4 + 2O_2 \rightarrow CO_2 + 2H_2O$ **12.** $3Ni + 2As \rightarrow Ni_3As_2$

13. $2Al + N_2 \rightarrow 2AlN$ **14.** $2C_2H_2 + 5O_2 \rightarrow 4CO_2 + 2H_2O$

15. $3Mg + Fe_2O_3 \rightarrow 3MgO + 2Fe$ **16.** $4K + O_2 \rightarrow 2K_2O$

17. $Sn + Br_2 \rightarrow SnBr_2$ **18.** $3Mg + N_2 \rightarrow Mg_3N_2$

19. $2H_2S + 3O_2 \rightarrow 2SO_2 + 2H_2O$

20. $Ca(OH)_2 + Na_2CO3 \rightarrow 2NaOH + CaCO_3$

21. $2FeCl_3$ + $SnCl_2$ $\longrightarrow$ $2FeCl_2$ + $SnCl_4$

　　　3+　　　　2+　　　　　　　2+　　　　4+

　　　└── 2 × (1e gained per Fe) ┘

　　　　　　　　　　── 2e lost ──

22. $3H_2SO_3$ + $2HNO_3$ $\longrightarrow$ $3H_2SO_4$ + $2NO$ + H_2O

　　　4+　　　　5+　　　　　　6+　　　2+

　　　└── 3 × (2e lost per S) ──┘

　　　　　└── 2 × (3e gained per N) ┘

23. $2KMnO_4$ + $5KNO_2$ + $3H_2SO_4 \rightarrow 2MnSO_4 + 3H_2O + 5KNO_3 + K_2SO_4$

　　　7+　　　　3+　　　　　　　2+　　　　　　　5+

　　　└ 2 × (5e gained per Mn) ──┘

　　　　　└── 5 × (2e lost per N) ────┘

24. $10HCl + 6HCl$ + $2KMnO_4 \rightarrow 8H_2O$ + $2KCl$ + $2MnCl_2$ + $5Cl_2$

　　　1-　　　　　　7+　　　　　　　　　　　2+　　　　0

　　　└ 5 × (2e lost per Cl_2) ──────┘

　　　　　└── 2 × (5e lost per Mn) ──┘

68

25. MnO_2 + 2HCl + 2HCl $\longrightarrow$ 2H$_2$O + MnCl$_2$ + Cl$_2$
   4+   1−   2+   0

   └─── 2e gained │ per Mn ─────────────────┘
                  └──────── 2e lost per Cl$_2$ ──────────┘

26. H_2S + 3/2O$_2$ $\longrightarrow$ S O$_2$ + H$_2$O
   2−   0   4+2−   2−

   └─ 6e lost │per S ──────────────┘ │
             └─ 2e gained per O ───┘ └─────────────┘

   2H$_2$S + 3O$_2$ → 2SO$_2$ + 2H$_2$O

27. a **28.** c **29.** b **30.** b

31. d, $2.80 \text{ mol} \times \dfrac{9.01 \text{ g}}{1 \text{ mol}} = 25.2$ **32.** a, $0.360 \times \dfrac{40.0 \text{ g}}{1 \text{ mol}} = 14.4 \text{ g}$

33. c, $3 \times 10^{22} \text{ atoms} \times \dfrac{1 \text{ mol}}{6 \times 10^{23} \text{ atoms}} \times \dfrac{4.00 \text{ g}}{1 \text{ mol He}} = 0.2 \text{ g}$

34. d **35.** c, $(2 \times 27.0) + 3[12.0 + (3 \times 16.0)] = 234 \text{ amu}$

36. a **37.** (a) 86.9 amu (b) 213 amu (c) 44.0 amu (d) 217 amu
(e) 123 amu (f) 98.1 amu (g) 2.00 amu (h) 103 amu

38. (a) $3.00 \text{ mol} \times \dfrac{86.9 \text{ g}}{1 \text{ mol MnO}_2} = 261 \text{ g}$

(b) $2.00 \times 10^{21} \text{ atoms} \times \dfrac{1 \text{ mol}}{6.02 \times 10^{23} \text{ atoms}} \times \dfrac{213 \text{ g}}{1 \text{ mol}} = 0.708 \text{ g}$

(c) $0.250 \text{ mol} \times \dfrac{44.0 \text{ g}}{1 \text{ mol}} = 11.0 \text{ g}$ (d) $4.50 \text{ mol} \times \dfrac{217 \text{ g}}{1 \text{ mol}} = 976 \text{ g}$

39. (a) $2.40 \times 10^{24} \times \dfrac{1 \text{ mol}}{6.02 \times 10^{23} \text{ atoms}} = 4.00 \text{ mol}$

(b) $0.482 \text{ g} \times \dfrac{1 \text{ mol}}{98.1 \text{ g}} = 0.00491 \text{ mol} = 4.91 \times 10^{-3} \text{ mol}$

(c) $1.00 \text{ kg} \times \dfrac{1000 \text{ g}}{1 \text{ kg}} \times \dfrac{1 \text{ mol}}{2.00 \text{ g}} = 500 \text{ mol}$

(d) $4.60 \text{ mg} \times \dfrac{1 \text{ g}}{1000 \text{ mg}} \times \dfrac{1 \text{ mol}}{103 \text{ g}} = 4.46 \times 10^{-5} \text{ mol}$

40. (a) One molecule of nitrogen reacts with three molecules of hydrogen to produce two molecules of ammonia. (b) One mole of nitrogen reacts with three moles of hydrogen to produce two moles of ammonia. (c) 28.0 grams of nitrogen react with 6.00 grams of hydrogen to produce 34.0 grams of ammonia. **41.** (a) One formula unit of iron(III) chloride reacts with three formula units of sodium hydroxide to form one formula unit of iron(III) hydroxide and three formula units of sodium chloride. (b) One mole of iron(III) chloride reacts with three moles of sodium hydroxide to form one mole of iron(III) hydroxide and three moles of sodium chloride. (c) 162 grams of iron(III) chloride reacts with 120 grams of sodium hydroxide to form 107 grams of iron(III) hydroxide and 176 grams of sodium chloride. **42.** (a) One molecule of fluorine reacts with two formula units of sodium chloride to produce two formula units of sodium fluoride and one molecule of chlorine. (b) One mole of fluorine will react with two moles of sodium chloride to produce two moles of sodium fluoride and one mole of chlorine. (c) 38.0 grams of fluorine will react with 117 grams of sodium chloride to produce 84.0 grams of sodium fluoride and 71.0 grams of chlorine.

43. $4NH_3 + 3O_2 \rightarrow 2N_2 + 6H_2O$

$$51.0 \text{ g } NH_3 \times \frac{1 \text{ mol } NH_3}{17.0 \text{ g}} \times \frac{6 \text{ mol } H_2O}{4 \text{ mol } NH_3} \times \frac{18.0 \text{ g}}{1 \text{ mol } H_2O} = 81.0 \text{ g}$$

44. $2NaCl \rightarrow 2Na + Cl_2$

$$2.00 \text{ tons} \times \frac{1000 \text{ kg}}{1 \text{ ton}} \times \frac{1 \text{ mol } Cl_2}{0.0710 \text{ kg}} \times \frac{2 \text{ mol } NaCl}{1 \text{ mol } Cl_2} \times \frac{0.0585 \text{ kg}}{1 \text{ mol } NaCl}$$

$$= 3.30 \times 10^3 \text{ kg}$$

45.

(a) $$27.0 \text{ g } Ag \times \frac{1 \text{ mol } Ag}{108 \text{ g}} \times \frac{2 \text{ mol } HNO_3}{1 \text{ mol } Ag} \times \frac{63.0 \text{ g}}{1 \text{ mol } HNO_3} = 31.5 \text{ g}$$

(b) $$27.0 \text{ g } Ag \times \frac{1 \text{ mol } Ag}{108 \text{ g}} \times \frac{1 \text{ mol } NO_2}{1 \text{ mol } Ag} \times \frac{46.0 \text{ g}}{1 \text{ mol } NO_2} = 11.5 \text{ g}$$

(c) $$27.0 \text{ g } Ag \times \frac{1 \text{ mol } Ag}{108 \text{ g}} \times \frac{1 \text{ mol } AgNO_3}{1 \text{ mol } Ag} \times \frac{170 \text{ g}}{1 \text{ mol } AgNO_3} = 42.5 \text{ g}$$

46. $N_2 + 3H_2 \rightarrow 2NH_3$

$$1 \text{ ton } N_2 \times \frac{2000 \text{ lb}}{1 \text{ ton}} \times \frac{454 \text{ g}}{1 \text{ lb}} \times \frac{1 \text{ mol } N_2}{28.0 \text{ g}} = 3.24 \times 10^4 \text{ mol } N_2$$

$$0.500 \text{ ton } H_2 \times \frac{2000 \text{ lb}}{1 \text{ ton}} \times \frac{454 \text{ g}}{1 \text{ lb}} \times \frac{1 \text{ mol } H_2}{2.02 \text{ g}} = 22.5 \times 10^4 \text{ mol } H_2$$

$$3.24 \times 10^4 \text{ mol } N_2 \times \frac{3 \text{ mol } H_2}{1 \text{ mol } N_2} = 9.72 \times 10^4 \text{ mol } H_2$$

Nitrogen is the limiting reagent:

$$3.24 \times 10^4 \text{ mol } N_2 \times \frac{2 \text{ mol } NH_3}{1 \text{ mol } N_2} \times \frac{17.0 \text{ g}}{1 \text{ mol } NH_3} \times \frac{1 \text{ lb}}{454 \text{ g}} \times \frac{1 \text{ ton}}{2000 \text{ lb}}$$

$$= 1.21 \text{ ton } NH_3$$

Chapter 6 THE THREE STATES OF MATTER

In this chapter, we discuss the three states in which matter can exist: solid, liquid, and gas. We examine how they differ and how they can be converted one to another. We also discuss the laws that describe the behavior of matter in the gaseous state.

6.1 Solids

A solid has a definite shape and a definite volume. Its particles are close together and arranged either in a disordered fashion (as in an amorphous solid), or in a highly ordered fashion (as in a crystalline solid.) Solid particles have restricted motion and can only vibrate around a fixed point. As a solid is heated, the particles vibrate more and more violently. Finally, a temperature is reached at which the structure of the solid begins to break down. This temperature is the melting point of the solid. The heat of fusion is the amount of energy required to change one gram of a solid into a liquid at the melting point.

Important Terms

solid	amorphous solid	crystalline solid
melting point	heat of fusion	

Example _____

How many kilocalories of energy are required to melt 150 grams of benzene if the heat of fusion of benzene is 30 cal/g?

$$150 \text{ g} \times \frac{30 \text{ cal}}{1 \text{ g}} \times \frac{1 \text{ kcal}}{1000 \text{ cal}} = 4.5 \text{ kcal}$$

6.2 Liquids

A liquid has a definite volume but not a definite shape; it will take on the shape of the container. Particles in a liquid are close together, but are not held in place as tightly as in a solid and are able to slide over one another. Two properties of a liquid resulting from the attraction between the liquid particles are viscosity (a measure of how easily a liquid flows)

particles are viscosity (a measure of how easily a liquid flows) and surface tension (the resistance of the particles on the surface of a liquid to the expansion of that liquid). Very energetic particles near the surface of a liquid can escape to the gaseous state in a process called evaporation. The opposite of evaporation is condensation: the formation of liquid as gas particles are cooled. Sublimation is a change of state in which a solid goes directly into the gaseous state.

As heat is added to a liquid, the particles move more and more rapidly until a temperature is reached at which bubbles of vapor are formed within the liquid. Under conditions of normal atmospheric pressure, this temperature is known as the normal boiling point of the liquid. The heat of vaporization of a liquid is the amount of energy required to change one gram of a liquid into a gas at the boiling point of the liquid.

Important Terms

liquid	viscosity	surface tension
evaporation	condensation	boiling point
sublimation	heat of vaporization	

Example ———

How many kilocalories are released when 33 grams of ethanol are condensed (converted from a gas to a liquid) if the heat of vaporization of ethanol is 200 cal/g?

$$33 \text{ g} \times \frac{200 \text{ cal}}{1 \text{ g}} \times \frac{1 \text{ kcal}}{1000 \text{ cal}} = 6.6 \text{ kcal}$$

Self-Test ———

For questions 1-12, indicate whether the statement is true (T) or false (F).

F 1.   Particles of a substance in the solid state have more kinetic energy than particles of that substance in the liquid state.

T 2.   Particles in the gaseous state are much more widely separated than particles in the liquid state.

T 3.   At normal atmospheric pressure, the temperature of boiling water will remain at 100°C no matter how much heat energy is added to the water.

F 4.   Because liquid particles are held in a fixed position, a liquid will take on the shape of the container.

72

F 5. A diamond is an example of an amorphous solid.

F 6. Oil is less viscous than water.

T 7. The higher the surface tension of a liquid, the more the particles on the surface of the liquid will prevent the liquid from spreading out.

T 8. Adding turpentine to some paints will make the paint less viscous.

f 9. The temperature at which a liquid becomes a gas is its melting point.

f 10. When particles of a liquid evaporate, the average kinetic energy of the liquid increases.

F 11. Lowering the temperature of a gas increases the kinetic energy of the gas particles.

T 12. A real gas does not behave like an ideal gas when the temperature of the gas is very low.

13. Indicate whether each of the following processes are endothermic or exothermic.
   (a) boiling EN (d) condensation EX
   (b) sublimation EN (e) melting EN
   (c) freezing EX (f) evaporation EN

14. How much heat energy is required to convert 1 liter of water at 100°C to steam at 100°C? (Remember that the density of water = 1 g/mL.)

$$1000\,mL \times \frac{1g}{1\,mL} \times \frac{539\,cal}{1g} = 539,000\,cal = \frac{1\,kcal}{1000\,cal} = 539\,kcal$$

6.3   Units of Pressure

The pressure exerted by a gas results from the collisions of the gas particles against the sides of the container. The definition of pressure is force per unit area. Pressure can be measured in many units. You should know the definition of an atmosphere, millimeters of mercury, torr, and pascal.

Important Terms

atmosphere (atm)　millimeters of mercury (mm Hg)
torr (torr)　standard atmospheric pressure
pascal (Pa)　standard temperature and pressure (STP)

73

For questions 15 to 20, complete the conversions required.

15. 327 mm Hg = 0.430 atm   18. 1102 torr = 1.45 atm
16. 157 torr   = 59 mm Hg   19. 2.7 atm   = 2052 mm Hg
17. 0.850 atm = 646 torr   20. 56 torr   = 7464.8 Pa

                                              2100
                                              2520

6.4 Boyle's Law

Boyle's law describes the relationship between the volume and the pressure of a gas. If the pressure on the gas is increased, the gas will be squeezed into a smaller volume. Mathematically, Boyle's law states that at a constant temperature and number of moles, pressure times volume equals a constant (PV = constant). To solve problems using Boyle's law, we make use of the following equation:

$$P_1V_1 \;=\; P_2V_2 \quad \text{(at constant T and n)}$$

Important Term

    Boyle's law

1. If a gas occupies a volume of 750 mL at standard atmospheric pressure, what will be the volume of the gas if the pressure is increased to 950 mm Hg and the temperature remains constant?

    (a) We must first identify P_1, P_2, V_1, and V_2. (Note: Be certain the two pressures are expressed in the same units, and the two volumes are expressed in the same units.)

              P_1 = 760 mm Hg P_2 = 950 mm Hg
              V_1 = 750 mL V_2 = ?

    (b) Now ask yourself: will the starting volume of 750 mL increase or decrease if the pressure is increased from 760 mm Hg to 950 mm Hg? Boyle's law states that the volume will decrease if the pressure is increased. Therefore, we must multiply the volume by a ratio of pressures that will decrease the volume.

$$750 \text{ mL} \times \frac{760 \text{ mm Hg}}{950 \text{ mm Hg}} \;=\; 600 \text{ mL}$$

(c)   You can also solve this problem by using the equation for Boyle's law.

$$760 \text{ mm Hg} \times 750 \text{ mL} = 950 \text{ mm Hg} \times V_2$$

$$\frac{760 \text{ mm Hg} \times 750 \text{ mL}}{950 \text{ mm Hg}} = V_2$$

$$600 \text{ mL} = V_2$$

2.   A 25-liter sample of carbon dioxide that exerts a pressure of 760 torr has been compressed into a 5.0-liter cylinder. If the temperature remains constant, what is the pressure of the carbon dioxide in the cylinder?

Let's solve this problem following the same procedure as example 1.

(a) $P_1 = 760 \text{ torr}$ $P_2 = \ ?$
          $V_1 = 25 \text{ liters}$ $V_2 = \ 5.0 \text{ liters}$

(b)   If the volume of the gas decreases, the pressure will increase.

$$760 \text{ torr} \times \frac{25 \text{ liters}}{5.0 \text{ liters}} = 3800 \text{ torr}$$

(c) $760 \text{ torr} \times 25 \text{ liters} = P_2 \times 5.0 \text{ liters}$

$$\frac{760 \text{ torr} \times 25 \text{ liters}}{5.0 \text{ liters}} = P_2$$

$$3800 \text{ torr} = P_2$$

Self-Test

21.   A weather balloon is inflated to a volume of 200 liters on a day when the atmospheric pressure is 755 mm Hg. What would be the volume of the balloon at an altitude of 17,000 feet where the atmospheric pressure is 370 mm Hg? (Assume that the temperature remains constant.)

22.   A sample of oxygen is placed in a 600-mL cylinder at a pressure of 1.0 atm, and the volume of the cylinder is then compressed to 80 mL. What is the new pressure in the cylinder if the temperature remains constant?

23.   A gas occupies a volume of 280 mL at a pressure of 450 torr. If the gas is transferred to a 1.5-liter container, what is the new pressure (in atm) exerted by the gas?

6.5   Charles' Law

Charles' law describes the relationship between the volume of a gas and the temperature of the gas (in degrees Kelvin) when the pressure and number of moles of gas remain constant. If the temperature of a gas is increased, the volume of the gas will increase. Mathematically, we can express the relationship as follows:

$$\frac{V_1}{T_1} = \frac{V_2}{T_2} \qquad \text{(at constant P and n)}$$

Important Term

Charles' law

Example _____

1. A balloon contains 3.5 liters of helium at 30°C. What would be the volume of the helium if it is cooled to 10°C without changing the pressure?

    (a) First we must identify V_1, T_1, V_2, and T_2. (Remember that all temperatures must be expressed in degrees Kelvin.)

    $$V_1 = 3.5 \text{ liters} \qquad V_2 = \text{ ?}$$
    $$T_1 = 303 \text{ K } (30°C + 273) \qquad T_2 = 283 \text{ K}$$

    (b) Now ask yourself: will the volume increase or decrease if the temperature decreases from 30°C to 10°C? From Charles' law we know that the volume will decrease when the temperature decreases. So we must multiply the volume by a ratio of temperatures that will decrease the volume.

    $$3.5 \text{ liters} \times \frac{283 \text{ K}}{303 \text{ K}} = 3.3 \text{ liters}$$

    (c) Or we can solve the problem by substituting the values in the equation for Charles' law.

    $$\frac{3.5 \text{ liters}}{303 \text{ K}} = \frac{V_2}{283 \text{ K}}$$

    $$3.3 \text{ liters} = V_2$$

2. To what Celsius temperature must a gas be heated to triple its volume if it occupies 450 mL at 30°C? (The pressure is constant.)

(a)   Following the same steps as example 1:

$$V_1 = 450 \text{ mL} \qquad V_2 = 1350 \text{ mL}$$
$$T_1 = 303 \text{ K} \qquad T_2 = \quad ?$$

(b)   For the volume of a gas to increase while the pressure remains constant, the temperature must increase.

$$303 \text{ K} \times \frac{1350 \text{ mL}}{450 \text{ mL}} = 909 \text{ K or } 636°C$$

(c)

$$\frac{450 \text{ mL}}{303 \text{ K}} = \frac{1350 \text{ mL}}{T_2}$$

$$450 \text{ mL} \times T_2 = 1350 \text{ mL} \times 303 \text{ K}$$

$$T_2 = \frac{1350 \text{ mL} \times 303 \text{ K}}{450 \text{ mL}}$$

$$T_2 = 909 \text{ K or } 636°C$$

Self-Test _____

24.   A sample of helium gas occupies a volume of 250 mL at 37°C. What will be the new Celsius temperature of the hydrogen if it is condensed to one tenth its original volume? (The pressure remains constant.)

25.   A 2.0-liter sample of nitrogen gas is heated from 20°C to 90°c. What is the final volume of the nitrogen if the pressure remains constant?

6.6   Molar Volume

Avogadro's law states that under identical conditions of temperature and pressure, equal volumes of different gases will contain the same number of moles of gas. One mole of any gas will occupy 22.4 liters at standard temperature (0°C) and pressure (760 mm Hg).

Important Terms

Avogadro's law   molar volume

6.7   Ideal Gas Law

Charles' and Boyle's laws can be combined into a general gas law equation that is helpful in calculating the effects of changes in temperature, pressure, and volume on a gas. Mathematically,

$$\frac{P_1V_1}{T_1} = \frac{P_2V_2}{T_2} \text{ (at constant n)}$$

The ideal gas law takes into account changes in the number of moles of a gas.

$$PV = nRT \text{ (where R = 0.0821 liter atm/mol K)}$$

Important Terms

ideal gas law   universal gas constant

Example _____

1. A 11-liter sample of neon at 1.0 atm and −198°C expands to 22 liters at −173°C. What is the new pressure?

   (a) To solve this problem we identify all the variables:

   P_1 = 1.0 atm P_2 = ?
   V_1 = 11 liters V_2 = 22 liters
   T_1 = 75 K T_2 = 100 K

   (b) We can solve this problem by reasoning in two steps. First, the volume is increasing; from Boyle's law we know this should make the pressure decrease. Therefore, we multiply P_1 by a ratio of volumes that will decrease the pressure. Second, the temperature is increasing. The higher the temperature, the greater the kinetic energy of the gas particles. Therefore, the greater the pressure. So the ratio of temperatures to use is the one that will increase the pressure.

   $$1.0 \text{ atm} \times \frac{11 \text{ liters}}{22 \text{ liters}} \times \frac{100 \text{ K}}{75 \text{ K}} = 0.67 \text{ atm}$$

   (c) Substituting in the combined gas law equation:

   $$\frac{1.0 \text{ atm} \times 11 \text{ liters}}{75 \text{ K}} = \frac{P_2 \times 22 \text{ liters}}{100 \text{ K}}$$

   $$\frac{1.0 \text{ atm} \times 11 \text{ liters} \times 100 \text{ K}}{75 \text{ K} \times 22 \text{ liters}} = P_2$$

   $$0.67 \text{ atm} = P_2$$

2. If a sample of oxygen gas weighs 24.0 g at 25°C and 740 mm Hg, what volume would it occupy?

78

Because we are concerned with the amount of gas in this problem, we will use the ideal gas law. The first thing we must do is convert the given data into the units of the universal gas constant.

$$P = 740 \text{ mm Hg} \times \frac{1 \text{ atm}}{760 \text{ mm Hg}} = 0.974 \text{ atm}$$

$$T = 25°C + 273 = 298 \text{ K}$$

$$n = 24.0 \text{ g} \times \frac{1 \text{ mol}}{32.0 \text{ g}} = 0.75 \text{ mol } O_2$$

$$V = ?$$

Substituting in the ideal gas law equation, we have

$$0.974 \text{ atm} \times V = 0.75 \text{ mol} \times 0.0821 \frac{\text{liter atm}}{\text{mol K}} \times 298 \text{ K}$$

$$V = \frac{18.3 \text{ liter atm}}{0.974 \text{ atm}}$$

$$V = 18.8 \text{ liters}$$

Self-Test

26. A sample of carbon dioxide occupies a 0.500-liter container at 170 torr and 25°C.
    (a) What is the weight of this sample in milligrams?
    (b) What would be the pressure of this sample of carbon dioxide if it were placed in a 65.0-mL tube at 15°C? (Hint: make certain that the units for the initial and final P, V, and T are the same.)

27. A 3.50-liter cylinder was filled with butane gas to a pressure of 3650 torr at 20°C. The cylinder is made to withstand an internal pressure of 10.0 atm.
    (a) What is the maximum temperature to which the cylinder could be heated before exploding?
    (b) What would be the pressure in the cylinder if the amount of butane gas were tripled?

6.8 Graham's Law of Effusion

Graham's law of effusion states that the lighter the gas, the faster the rate of effusion. Effusion is the movement of gas particles through a tiny hole to a region of lower pressure. We can use the mathematical expression for Graham's law to compare

$$\frac{\text{Effusion rate of gas A}}{\text{Effusion rate of gas B}} = \sqrt{\frac{\text{formula weight of B}}{\text{formula weight of A}}}$$

Important Terms

    Graham's law   effusion   diffusion

Example _____

Compare the rate of effusion of oxygen gas with that of carbon dioxide.

$$\frac{\text{Effusion of } O_2}{\text{Effusion of } CO_2} = \sqrt{\frac{44}{32}} = \sqrt{1.375} = 1.2$$

Oxygen gas will effuse 1.2 times faster than carbon dioxide gas.

6.9 Henry's Law

Henry's law relates the pressure of a gas to its solubility in a liquid. This law states that the higher the pressure on a gas, the greater its solubility in a liquid (when the temperature remains constant).

Important Terms

    Henry's law   bends

6.10 Dalton's Law of Partial Pressures

Dalton's law states that the pressure exerted by a mixture of gases in a container is equal to the sum of the partial pressures of those gases. Each gas in the mixture acts independently of the others and will exert a pressure (called its partial pressure) as if it were alone in the container. The pressure exerted by a gas is directly related to the number of gas particles in the container. Gas collected over water will contain a mixture of gas and water molecules. The pressure of the water molecules is called the vapor pressure, and will be dependent upon the temperature of the water.

Important Terms

    Dalton's law   partial pressure   vapor pressure

6.11 The Diffusion of Respiratory Gases

The diffusion of respiratory gases in our bodies is directly
related to the partial pressures of those gases. The gases
(carbon dioxide and oxygen) will diffuse from a region of higher
partial pressure to a region of lower partial pressure.

Example

1. You have a 500-mL container of helium at a pressure of 450
    mm Hg, and a 500-mL container of oxygen at a pressure of 240
    mm Hg. You now place both samples in the same 500-mL
    container. What is the partial pressure of each gas, and
    what is the total pressure in the container?

    By the definition of partial pressure, P_{helium} = 450 mm Hg
    and P_{oxygen} = 240 mm Hg.

    Since the total pressure will equal the sum of the partial
    pressures,

    $$P_{total} = P_{helium} + P_{oxygen}$$

    $$= 450 \text{ mm Hg} + 240 \text{ mm Hg} = 690 \text{ mm Hg}$$

2. You collect a sample of hydrogen gas in a laboratory
    apparatus similar to Figure 6.14 in your text. The water
    temperature is 20°C and the barometer reads 738.5 mm Hg.
    The volume of the collected gas is 350 mL. Calculate the
    partial pressure of the hydrogen.

    Using Table 6.1 in your text, we see that the vapor pressure
    of water at 20°C is 17.5 torr. We know that 1 mm Hg = 1
    torr, so the atmospheric pressure = 738.5 mm Hg = 738.5
    torr.

    $$P_{oxygen} = P_{total} - P_{water\ vapor}$$

    $$= 738.5 \text{ torr} - 17.5 \text{ torr} = 721.0 \text{ torr}$$

6.12 The Kinetic Theory of Gases

The particles of a gas move very rapidly in a random chaotic
fashion. A gas has neither a definite volume nor a definite
shape, but will take on the volume and shape of a container. The
kinetic theory describing the properties of an ideal gas can be
summarized as follows:

1. The very small particles of a gas are widely separated from
    one another.

2. These particles are moving very fast in a random fashion and will travel in straight lines until they collide with one another or with the sides of the container.

3. A particle of gas doesn't lose any energy when it collides with another particle or with the sides of the container.

4. Gas particles are not attracted or repulsed by other gas particles.

5. The kinetic energy of the gas particles increases with an increase in temperature.

Real gases behave according to this theory except under conditions of very low temperatures or very high pressures.

Important Terms

gas   kinetic-molecular theory   absolute zero

Self-Test

For questions 28 to 36, choose the correct answer.

28. Pressure equals
    (a) millimeters of mercury   (c) force per unit area
    (b) volume/temperature   (d) temperature/volume

29. One atmosphere equals
    (a) 17.5 torr   (c) 1 mm Hg
    (b) 760 torr   (d) 133.3 Pa

30. Standard atmospheric pressure equals
    (a) 760 mm Hg   (c) 1 torr
    (b) 760 Pa   (d) 760 atm

31. If you decrease the volume of a gas when the temperature is constant, the pressure will
    (a) increase   (c) remain the same
    (b) decrease   (d) none of the above

32. If you increase the temperature of a gas when the pressure is constant, the volume will
    (a) increase   (c) remain the same
    (b) decrease   (d) none of the above

33. How many times faster will molecules of H_2 effuse than molecules of N_2.
    (a) 28   (c) 14
    (b) 0.27   (d) 3.7

34. If the pressure of a gas above a liquid is increased, its solubility in the liquid will
    (a) increase   (c) remain the same
    (b) decrease   (d) none of the above

35. The total pressure of a mixture of gases will equal
    (a) the atmospheric pressure minus the vapor pressure
    (b) the sum of the partial pressures of the gases
    (c) the sum of the partial pressures of the gases/volume
    (d) the volume time the partial pressures

36. Absolute zero equals
    (a) 0°C   (c) 0 K
    (b) -273 K   (d) 273°C

37. Which gas will effuse faster: ammonia or carbon dioxide? How much faster?

38. A gaseous mixture containing oxygen, nitrogen, carbon monoxide, and carbon dioxide has a pressure of 2.00 atm. What is the partial pressure of the oxygen if $P_{nitrogen}$ = 532 mm Hg, $P_{carbon\ monoxide}$ = 684 mm Hg and $P_{carbon\ dioxide}$ = 76 mm Hg?

39. A 2.0-liter sample of oxygen exerts a pressure of 480 torr, and a 2.0-liter sample of nitrogen exerts a pressure of 0.35 atm. What would be the total pressure if these two samples were combined in the same 2.0-liter container?

40. A flask contains chlorine gas and fluorine gas at a pressure of 1.10 atm. Are there a greater number of chlorine or fluorine gas molecules if $P_{chlorine}$ = 534 torr?

41. A student collected a sample of hydrogen gas in a 1-liter container by water displacement. In her notebook, she recorded the barometric pressure as 756.8 mm Hg and the water temperature as 25°C.
    (a) What is the partial pressure of the hydrogen gas that she collected?
    (b) If she recorded the volume of the collected gases as 350 mL, what volume would the dry hydrogen occupy at STP?

Answers to the Self-Test Questions in Chapter 6

(Answers rounded to the correct number of significant figures)

1. F 2. T 3. T 4. F 5. F 6. F 7. T 8. T 9. F 10. F
11. F 12. T 13. (a) endothermic (b) endothermic (c) exothermic
(d) exothermic (e) endothermic (f) endothermic

14. $1000 \text{ mL} \times \dfrac{1 \text{ g}}{1 \text{ mL}} \times \dfrac{539 \text{ cal}}{1 \text{ g}} = 539,000 \text{ cal} \times \dfrac{1 \text{ kcal}}{1000 \text{ cal}} = 539 \text{ kcal}$

15. 0.430 atm 16. 157 torr 17. 646 torr 18. 1.45 atm 19. 2100
mm Hg 20. 7500 Pa

21. $V_2 = \dfrac{200 \text{ liters} \times 755 \text{ mm Hg}}{370 \text{ mm Hg}} = 408 \text{ liters}$

22. $P_2 = \dfrac{600 \text{ mL} \times 1 \text{ atm}}{80 \text{ mL}} = 7.5 \text{ atm}$

23. $P_2 = \dfrac{280 \text{ mL} \times 450 \text{ torr}}{1500 \text{ mL}} = 84 \text{ torr} \times \dfrac{1 \text{ atm}}{760 \text{ torr}} = 0.11 \text{ atm}$

24. $T_2 = \dfrac{25 \text{ mL} \times 310 \text{ K}}{250 \text{ mL}}$ $T_2 = 31 \text{ K or } -242°C$

25. $V_2 = \dfrac{2.0 \text{ liters} \times 363 \text{ K}}{293 \text{ k}} = 2.5 \text{ liters}$

26. (a) $PV = nRT$: $P = 0.224$ atm, $V = 0.500$ liter, $n = ?$ mol
$R = 0.0821$ liter atm/mol K, $T = 298$ K

$n = \dfrac{PV}{RT} = \dfrac{0.224 \text{ atm} \times 0.500 \text{ liter}}{0.0821 \dfrac{\text{liter atm}}{\text{mol K}} \times 298 \text{ K}} = 0.00458 \text{ mol}$

$\text{mg } CO_2 = 0.00458 \text{ mol} \times \dfrac{44 \text{ g}}{1 \text{ mol}} \times \dfrac{1000 \text{ mg}}{1 \text{ g}} = 202 \text{ mg}$

(b) $\dfrac{170 \text{ torr} \times 500 \text{ mL}}{298 \text{ K}} = \dfrac{P_2 \times 65.0 \text{ mL}}{288 \text{ K}}$ $P_2 = 1260 \text{ torr}$

27. (a) $\dfrac{4.80 \text{ atm} \times 3.50 \text{ liters}}{293 \text{ K}} = \dfrac{10.0 \text{ atm} \times 3.50 \text{ liters}}{T_2}$

$T_2 = 610 \text{ K or } 337°C$

(b) $4.80 \text{ atm} \times 3.5 \text{ liter} = n \times 0.0821 \dfrac{\text{liter atm}}{\text{mol K}} \times 293 \text{ K}$

$n = 0.698 \text{ mol}$

$$P_2 \times 3.5 \text{ liter} = 2.09 \text{ mol} \times 0.0821 \text{ liter atm/mol K} \times 293 \text{ K}$$
$$P_2 = 14.4 \text{ atm}$$

28. c **29.** b **30.** a **31.** a **32.** a **33.** d **34.** a **35.** b **36.** c **37.** ammonia, 1.6 times **38.** 228 mm Hg **39.** P_T = 746 torr or 0.98 atm **40.** more chlorine, $P_{chlorine}$ = 534 torr, $P_{fluorine}$ = 302 torr **41.** (a) $P_{hydrogen}$ = 733.0 mm Hg

(b) $\dfrac{733 \text{ mm Hg} \times 350 \text{ mL}}{298 \text{ K}} = \dfrac{760 \text{ mm Hg} \times V_2}{273 \text{ K}}$ $\qquad V_2 = 309 \text{ mL}$

Chapter 7   THE ATOM AND
RADIOACTIVITY

In this chapter we focus our attention on the center of the
atom—the nucleus—and the changes that can take place within the
nucleus to make the atom more stable.

7.1   What is Radioactivity?

Some nuclei with a large ratio of positive protons to neutral
neutrons can be unstable, and can give off a particle to form a
more stable daughter nucleus. Elements or compounds that give
off these particles (or undergo decay) are said to be
radioactive. The daughter nucleus that is formed may also be
unstable, and the process of decay will continue until a stable
nucleus is formed. Such as sequence of decays is called a decay
series or disintegration series. Isotopes of an element that
give off radiation are called radionuclides or radioisotopes.

Important Terms

> radioactivity   disintegration series   decay series
> radionuclide   daughter nucleus   radioisotope

7.2   Alpha Radiation

There are several ways in which a nucleus can decay to become
more stable. One way is to give off an alpha particle, made up
of two protons and two neutrons (the nucleus of a helium atom.)
Although very small, an alpha particle is the largest of the
radioactive particles emitted by a decaying nucleus. Because of
its relatively large size, it has very little penetrating power.
The following is the nuclear equation for the decay of an
actinium-225 nucleus, which is an alpha emitter.

$$^{225}_{89}\text{Ac} \longrightarrow ^{221}_{87}\text{Fr} + ^{4}_{2}\text{He}$$

(Remember that the sum of the atomic numbers on each side of the
arrow must be equal, and the sum of the mass numbers on each side
of the arrow must be equal.)

Important Terms

> alpha radiation   alpha particle   nuclear equation

7.3 Beta Radiation

Beta radiation is composed of streams of electrons, produced within the nucleus and then given off in the process of decay. Because they are much smaller than alpha particles, beta particles have more penetrating power. The daughter nucleus produced in a beta decay will have the same mass number, but a different atomic number than the original nucleus. The following is an example of a beta decay:

$$^{66}_{28}\text{Ni} \longrightarrow {}^{66}_{29}\text{Cu} + {}^{0}_{-1}e$$

Important Terms

beta radiation   beta particle

7.4 Gamma Radiation

As we discussed in Chapter 2, gamma rays are a form of electromagnetic radiation having very short wavelength and very high energy. Gamma radiation has high penetrating power. Gamma rays are often given off in an alpha or beta decay as the daughter nucleus reaches a lower, more stable energy state. We indicate gamma radiation by including the Greek letter gamma (γ) in the nuclear equation. For example, the thorium-234 isotope is a beta and gamma emitter.

$$^{234}_{90}\text{Th} \longrightarrow {}^{234}_{91}\text{Pa} + {}^{0}_{-1}e + \gamma$$

Important Terms

gamma radiation   gamma rays

Self-Test _____

For questions 1 to 6, fill in the blank with the correct word or words.

1. A substance whose atoms give off particles to become more stable is said to be _____.

2. _____ radiation has more penetrating power than _____ radiation, but less penetrating power than _____ radiation.

3. _____ are nuclei of helium atoms.

4. Beta radiation consists of streams of _____.

5. _____ are a form of electromagnetic radiation with very _____ wavelengths.

6. Gamma rays have _____ charge; alpha particles have a _____ charge; and beta particles have a _____ charge.

7. Without looking back in the text or study guide, write the symbol for each of the following:
   (a) proton   (c) electron   (e) gamma ray
   (b) neutron   (d) alpha particle   (f) beta particle

For questions 8 to 13, fill in the missing symbol.

8. $^{148}_{64}Gd \longrightarrow {}^{144}_{62}Sm +$ _____

9. $^{90}_{38}Sr \longrightarrow {}^{90}_{39}Y +$ _____

10. $^{104}_{47}Ag \longrightarrow$ _____ $+ {}^{0}_{-1}e$

11. _____ $\longrightarrow {}^{241}_{95}Am + {}^{4}_{2}He$

12. $^{210}_{84}Po \longrightarrow$ _____ $+ {}^{4}_{2}He$

13. _____ $\longrightarrow {}^{73}_{32}Ge + {}^{0}_{-1}e$

7.5   Half-Life

Each radioactive substance decays at its own characteristic rate. The rate of decay is measured by the half-life of the radionuclide. The half-life is the amount of time it takes for one-half of the atoms in a sample to decay. The more unstable the nucleus, the more rapidly it will decay and the shorter the half-life. The half-lives of radioactive substances provide a useful tool for establishing the age of human artifacts, once-living organisms, and geologic periods.

Important Terms

half-life ($t_{1/2}$)   radioisotopic dating

Self-Test _____

14. The half-life of fermium-253 is 4.5 days. How many grams of a 2.0-gram sample would remain after 13.5 days?

15. Iodine-128 is used in medical diagnosis and has a half-life of 25 minutes. What fraction of the original sample would remain after 2 1/2 hours?

7.6 Nuclear Transmutation

Another way in which nuclei can undergo change is by being bombarded with high-speed particles. A high-speed particle can collide with a nucleus, producing a new nucleus by the process called nuclear transmutation. Nuclear transmutation is used in laboratories to produce new elements.

Important Term

nuclear transmutation

Self-Test

For questions 16 to 18, complete the transmutations.

16. $^{96}_{42}Mo$ + _____ $\longrightarrow$ $^{1}_{0}n$ + $^{97}_{43}Tc$

17. $^{239}_{94}Pu$ + $^{4}_{2}He$ $\longrightarrow$ _____ + $^{1}_{0}n$

18. _____ + $^{1}_{1}H$ $\longrightarrow$ $^{64}_{30}Zn$

7.7 Nuclear Fission

To meet the energy needs of our society, we are harnessing the energy produced by nuclear reactions. The nuclei of certain isotopes can be broken apart to form smaller, more stable nuclei. This process, called fission, occurs when one of these isotopes is struck by a neutron. The isotope then breaks apart to form two smaller nuclei, two or three neutrons, and a tremendous amount of energy. The neutrons so produced can react with other fissionable nuclei to cause further reactions. If a critical mass of fuel nuclei is present, this process can lead to a chain reaction

Important Terms

fission   chain reaction   critical mass

7.8 Nuclear Reactors

The energy generated by nuclear fission can be harnessed in a constructive manner in a nuclear reactor. The design of the reactor depends upon its use, but the basic components of the reactor core are the same: the fuel, a moderator, control rods, a heat transfer fluid, and shielding. In current design of nuclear reactors, the uranium-235 isotope is constantly being consumed. Since the world's natural supply of uranium-235 is small, the possibility long-term use of nuclear power depends on breeder reactors--reactors in which more fuel material is produced than is consumed. The critical factor in the development of breeder reactors is safety, including the handling and transport of radioactive materials and the safe disposal of radioactive wastes.

Important Terms

nuclear reactor   breeder reactor

7.9   Nuclear Wastes

The radioactive wastes produced in the fission process have long half-lives and must be stored safely for long periods of time. These wastes, which are found in spent fuel from nuclear power plants and from nuclear-related defense activities, are classified as high-level wastes. Other sources of radioactive wastes are mill tailings, waste from medical and commercial processes, and transuranic waste. High-level wastes are accumulating in temporary storage facilities around the country. The U.S. Department of Energy is responsible for establishing a permanent storage facility for high-level wastes, but location of this site is still being debated.

7.10   Nuclear Fusion—A Captured Sun

Small nuclei can also react by combining with one another to form heavier, more stable nuclei. This process, called nuclear fusion, yields a tremendous amount of energy. Nuclear fusion reactions generate the energy given off by our sun. If we could develop the technology necessary to harness nuclear fusion reactions, this process would have great advantages over nuclear fission. Delay in developing this technology is caused by the fact that nuclear fusion reactions require temperatures of one hundred million degrees Celsius or more to occur.

Important Terms

nuclear fusion   deuterium   tritium

Self-Test _____

For questions 19 to 28, choose the best answer or answers from the choices given.

19.   This reaction involves the joining of smaller nuclei to form a larger, more stable nucleus.
  (a) nuclear fusion   (c) nuclear transmutation
  (b) nuclear fission   (d) beta decay

20. This reaction involves the splitting apart of a large nucleus into two smaller, more stable nuclei.
    (a) nuclear fusion   (c) nuclear transmutation
    (b) nuclear fission   (d) beta decay

21. A requirement for nuclear fusion to occur is
    (a) a critical mass of uranium-235
    (b) a moderator
    (c) control rods
    (d) very high temperatures

22. A nuclear chain reaction can occur only when
    (a) the temperature is very high
    (b) a moderator is present.
    (c) a critical mass of uranium-235 is present
    (d) a heat transfer fluid is present

23. This removes heat from the reactor core in a nuclear reactor.
    (a) the shielding   (c) the control rods
    (b) the heat transfer fluid (d) the moderator

24. This functions to slow down the neutrons produced in a nuclear reactor.
    (a) the shielding   (c) the control rods
    (b) the heat transfer fluid (d) the moderator

25. This protects the surrounding environment from nuclear radiation.
    (a) the shielding   (c) the control rods
    (b) the heat transfer fluid (d) the moderator

26. This may contain plutonium-239 in a nuclear reactor.
    (a) the core   (c) the control rods
    (b) the fuel   (d) the moderator

27. This absorbs the neutrons produced by the fission reactions in a nuclear reactor.
    (a) the fuel   (c) the control rods
    (b) the heat transfer fluid (d) the moderator

28. Which of the following are advantages of nuclear fusion over nuclear fission?
    (a) It is more efficient.
    (b) It doesn't require high temperatures.
    (c) The fuel is abundant.
    (d) There is little radioactive waste.
    (e) There is little chance of the reactor core melting.
    (f) All of the above.

1. radioactive 2. beta, alpha, gamma 3. alpha particles
4. electrons 5. gamma rays, short 6. no, positive, negative

7.   (a) ^1_1p (b) ^1_0n (c) $^0_{-1}e$ (d) ^4_2He (e) γ (f) $^0_{-1}e$

8. ^4_2He 9. $^0_{-1}e$ 10. $^{104}_{48}Cd$ 11. $^{245}_{97}Bk$ 12. $^{206}_{82}Pb$ 13. $^{73}_{31}Ga$

14.   13.5 days $\times \dfrac{1\ \text{half-life}}{4.5\ \text{days}} = 3$ half-lives

   grams remaining = 2.0 g $\times (\tfrac{1}{2})^3 = 0.025$ g

15.   2.5 hr $\times \dfrac{60\ \text{min}}{1\ \text{hr}} \times \dfrac{1\ \text{half-life}}{25\ \text{min}} = 6$ half-lives

   $(\tfrac{1}{2})^6$ or 1/64 of the sample will remain

16. ^2_1H 17. $^{242}_{96}Cm$ 18. $^{63}_{29}Cu$

19. a 20. b 21. d 22. c 23. b 24. d 25. a 26. b 27. c
28.   a,c,d,e

Chapter 8   RADIOACTIVITY AND THE LIVING ORGANISM

In Chapter 7 we saw that unstable nuclei could become more stable by giving off various forms of radiation. In this chapter we examine the effects of radiation on living organisms, some methods for measuring this radiation, and the ways in which radiation can be used to benefit humans.

8.1   Ionizing Radiation

Radiation interacts with living cells to produce highly reactive particles. These particles may be charged particles called ions, or may be high-energy, uncharged particles called free radicals. Both ions and free radicals produce harmful changes in living cells. Such changes come about in two ways: by direct action or by indirect action. The radiation can interact directly with a biologically important molecule and damage it, or can interact with water in a cell to form ions or free radicals that then cause damage. Cells have efficient repair mechanisms to protect against damage to DNA, and contain molecules such as vitamin C and A that quickly neutralize any ions or free radicals that are formed. Cells with changed or altered DNA are called mutant cells. If such mutant cells divide and grow in an uncontrolled fashion, they are called cancerous or malignant.

Important Terms

ionizing radiation	ion	free radical
malignant cell	mutant cell	cancerous cell

8.2   Radiation Dosage

This section discusses various units used to measure ionizing radiation, or the effects of the radiation on living tissues. The curie measures the rate of radioactive emissions from a radioactive source. The rem and the sievert indicate absorbed dosage of radiation independent of the type of radiation used. Ionizing radiations interact differently with living tissue. Alpha particles leave short, dense clusters of ions, while gamma particles tend to leave scattered ion pairs along a much longer path.

Important Terms

curie (Ci)	rem	sievert (Sv)

8.3 Detecting Radiation: Radiation Dosimetry

Because ionizing radiation is potentially so dangerous to living organisms, it is important to monitor the level of exposure to such radiation. This section discusses some of the instruments used to measure ionizing radiation. One of the most widely used instruments is the Geiger-Müller counter, which is sensitive to beta radiation and measures the number of radiations given off (but not the energy of those radiations). The scintillation counter, on the other hand, is able to record both the number of radiations and the dose rate. Individuals who work closely with ionizing radiation wear small dosimeters, such as film badges.

Important Terms

dosimetry   Geiger-Müller counter   scintillation counter
film badge   dosimeter

8.4 Protection Against Radiation Exposure

To protect yourself from the damaging effects of ionizing radiation, you should minimize your exposure to it, remain as far away from the source of radiation as possible, and use proper shielding.

Important Terms

shielding   inverse square law

8.5 Background Radiation

Each of us is exposed to a small amount of ionizing radiation each day from natural sources. This radiation is called background radiation, and is produced by naturally occurring radionuclides that may be in the soil, food, water, or air. The average person receives only a small dose of radiation (about 400 rems) per year.

Important Term

background radiation

Self-Test _____

For questions 1 to 3, fill in the blank with the correct word or words.

1. Radiation causes harmful effects in living tissues by producing highly reactive particles called _____ and _____.

2. Ionizing radiation can produce its harmful effects in two ways:
   (a) By _____, in which the radiation interacts with water molecules to produce reactive particles that damage the important molecules in the cell. _____ and _____ radiation can produce this type of cellular damage.
   (b) By _____, in which the radiation reacts with a biologically important molecule, producing useless fragments. _____ and _____ radiation can produce this type of cellular damage.

3. Radiation sickness often resembles poisoning by the chemical _____.

4. Match each of the following definitions with the correct unit or units for measuring radiation. (Use Table 8.1)

   (a) Produces the same biological effect as one rad of therapeutic X rays

   (b) Measures only the frequency of disintegrations

   (c) Measures a specific amount of energy released per gram of irradiated tissue

   (d) Measures the amount of energy transferred to the irradiated tissue in a specified distance traveled by the radiation.

   (e) The SI unit that equals 100 rem.

   A. LET
   B. becquerel
   C. rem
   D. rad
   E. curie
   F. gray
   G. sievert

For questions 5 to 15, indicate whether the statement is true (T) or false (F).

5. A Geiger-Müller counter is most sensitive to gamma radiation.

6. A film badge is a good way to monitor the radiation exposure of an X-ray technician.

7. You would use a Geiger-Müller counter to measure the energy of the radiation given off by a sample of radioactive material.

8. Most of the internal radiation exposure of an individual comes from inhaled radon.

9. The average person is exposed to more ionizing radiation from the natural environment than from any other source.

10. The biological effects of one rad of various types of radiation will be the same.

11. The cell has mechanisms to repair the damage caused by ionizing radiation.

12. Cells with altered DNA are called mutant cells.

13. You would receive a more intense dose of radiation from a sample of cobalt-60 standing 9 feet from the sample than 3 feet from it.

14. Alpha particles are highly penetration ionizing radiation.

15. A cell is more likely to be able to repair the damage caused by gamma radiation than alpha radiation.

8.6   Medical Diagnosis

Radioactive chemicals called tracers have become increasingly useful in medical diagnosis. Both natural and artificially produced radionuclides are used. Compounds containing radioactive elements are synthesized and used to examine a specific area of the body (such as the thyroid or brain), or a specific function in the body (such as blood flow or movement of bile). With the development of highly sensitive measuring devices, diagnostic pictures can be produced using radionuclides that expose the patient to much lower dosages. A CT scanner uses X rays to produce detailed pictures of cross-sections of the human body, eliminating the need for many invasive diagnostic tests and exploratory surgeries. PET scans record gamma radiation produced when positrons given off by a radioactive tracer interact with electrons in the body. These scans produce important information about brain function and the development of disease.

Important Terms

tracer   gamma camera   CT scanner
PET scan

8.7   Radionuclides Used in Diagnosis

Technetium-99^m is a gamma emitter with a half-life of six hours. It is used extensively in the study of the liver, spleen, heart and bone marrow. Because of its desirable properties, technetium-99^m is replacing most other isotopes in common diagnostic procedures.

Important Terms

metastable   metastasize

8.8   Radiation Therapy

Ionizing radiation can be used to treat cancerous tissue because cancer cells divide more rapidly than normal cells and are more sensitive to the effects of ionizing radiation. High-intensity radiation from an X-ray machine, a cobalt-60 source, or a particle accelerator is used to treat cancerous tissue that can't be removed by surgery. To treat certain other cancers, radionuclides are inserted into the tumor with a needle. Scientists have also synthesized radionuclide-containing chemicals that will concentrate in the region or tissue that is to be treated.

Self-Test _____

For questions 16 to 26, choose the best answer or answers.

16.   This is effective in locating and diagnosing brain tumors.
      (a) gamma camera scan   (c) PET scan
      (b) CT scan   (d) MRI scan

17.   This isotope is used to treat thyroid cancer.
      (a) iodine-131   (c) strontium-90
      (b) radium-226   (d) phosphorus-32

18.   This isotope acts like potassium and concentrates in undamaged heart muscle.
      (a) iodine-131   (c) thallium-201
      (b) radium-226   (d) phosphorus-32

19.   Instrument used to produce a picture showing the location of a radioactive tracer.
      (a) gamma camera   (c) PET scanner
      (b) CT scanner   (d) MRI scanner

20.   Instrument used to follow the metabolic activity of the brain.
      (a) gamma camera   (c) PET scanner
      (b) CT scanner   (d) MRI scanner

21.   Instrument that produces detailed cross-sectional images of the body.
      (a) gamma camera   (c) PET scanner
      (b) CT scanner   (d) MRI scanner

22. This isotope is inserted by needle to slow tumor growth.
    (a) thallium-201   (c) technetium-99^m
    (b) radium-226   (d) gold-198

23. When cancer cells spread from the original tumor, they have
    (a) translated   (c) metastasized
    (b) transformed   (d) replicated

24. This isotope is in an energy state that is higher than normal
    (a) thallium-201   (c) technetium-99^m
    (b) radium-226   (d) gold-198

25. This isotope is used to determine if cancer cells from a tumor have spread to the bones.
    (a) phosphorus-32   (c) strontium-90
    (b) technetium-99^m (d) radium-226

26. Chemicals that contain radioactive atoms and that can be used to monitor living functions are called
    (a) metastable   (c) tracers
    (b) isotopes   (d) radionuclides

For questions 27 to 33, fill in the blank with the correct word or words.

27. Three advantages of technetium-99^m over other radioactive tracers are _____, _____, and _____.

28. A _____ analyzes the chemical composition of a tissue to show early stages of a disease.

29. A CT scanner forms its image using _____.

30. The isotope technetium-99^m is in the _____ state and releases _____ to become more stable.

31. Cancerous tissue that can't be removed by surgery might be treated by _____, _____, and _____.

32. Cells that divide _____(slowly, rapidly) are the most sensitive to ionizing radiation.

33. A radiologist selecting a radioisotope for use in diagnosis would try to find one that
    (a) Has a very long half-life, or has a short half-life
    (b) Is quickly eliminated, or is retained by the body
    (c) Will give results using 10 mg, or will give results using 50 mg

98

Answers to Self-Test Questions in Chapter 8

1. ions, free radicals **2.** (a) indirect action; beta, gamma,
X rays (b) direct action; alpha particles, neutrons **3.** hydrogen
peroxide **4.** (a) C (b) B,E (c) D,F (d) A (e) G **5.** F **6.** T **7.** F
8. T **9.** T **10.** F **11.** T **12.** T **13.** F **14.** F **15.** T **16.** b **17.** a
18. c **19.** a,c **20.** c **21.** b **22.** d **23.** c **24.** c **25.** b **26.** c
27. short half-life, gives off only gamma radiation, and the
energy of the gamma radiation can be easily detected **28.** PET
scan **29.** X rays **30.** metastable, gamma radiation
31. irradiation with X rays, insertions of a radionuclide into
the tumor, or treatment with a chemical containing a radionuclide
32. rapidly **33.** (a) short half-life (b) quickly eliminated (c)
will give results using 10 mg

Chapter 9   REACTION RATES AND CHEMICAL EQUILIBRIUM

Why do chemical reactions proceed at different rates? Why do some reactions give off heat, while others cannot occur unless heat is added? What factors can change the rate of a chemical reaction? Why do some chemical reactions proceed for a while and then seem to stop? These are a few of the questions about chemical reactions that are answered in this chapter. To discuss reaction rates and chemical equilibrium, we must introduce some new vocabulary, so be sure that you understand the important terms for each section.

9.1   Activation Energy

For a chemical reaction to occur, the reactant particles must collide with enough energy to overcome the forces of repulsion between the electrons surrounding each particle. In addition, the particles must collide with enough energy to break old bonds so that new bonds can be formed. The amount of energy necessary for a collision between reactant particles to result in the formation of products is called the activation energy. Each reaction has a characteristic activation energy. The higher the activation energy, the greater the energy the reactant particles must possess to collide successfully and form products.

Important Terms

> activation energy   activated complex

9.2   Exothermic and Endothermic Reactions

Chemical reactions can be classified as exothermic (those that give off energy) or endothermic (those that require energy to occur). The heat of reaction, ΔH, is the amount of energy given off or required in a reaction. The value of the heat of reaction is equal to the difference between the potential energy of the products and the potential energy of the reactants. The relationship between the potential energy of the reactants and products can be shown on a potential energy diagram. Study carefully the diagrams for an exothermic and endothermic reaction shown in Figure 9.5 in the text. From these diagrams you can see that the value of ΔH will be negative if the reaction is exothermic, and positive if the reaction is endothermic.

Important Terms

endothermic reaction   exothermic reaction
potential energy diagram   heat of reaction (ΔH)

Self-Test _____

For questions 1 to 5, answer true (T) or false (F).

F 1.   The heat of reaction is the minimum energy required in a
        collision between reactant particles for a reaction to take
        place.

F 2.   An endothermic reaction gives off energy.

T 3.   The products will have less potential energy than the
        reactants in an exothermic reaction.

F 4.   The activation energy of a reaction is the difference
        between the potential energy of the products and the
        potential energy of the reactants.

T 5.   If the activation energy barrier is high, most collisions
        between reactant molecules at room temperature will not be
        successful.

6.   Identify each of the following reactions as exothermic or
        endothermic.

        (a) $2H_2$ + O_2 $\longrightarrow$ $2H_2O$ + 115.6 kcal *Ex*
        (b) H_2 + I_2 + 12.4 kcal $\longrightarrow$ 2HI *En*
        (c) SO_2 + 71 kcal $\longrightarrow$ S + O_2 *En*
        (d)   2C + $3H_2$ $\longrightarrow$ C_2H_6 + 20.2 kcal *Ex*

7.   The following is the potential energy curve for the reaction

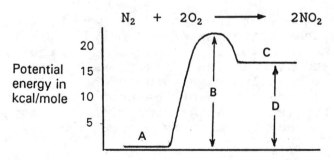

N_2 + $2O_2$ $\longrightarrow$ $2NO_2$

        (a)   What letter represents $2NO_2$? C
        (b)   What letter represents N_2 + $2O_2$? A
        (c)   What letter represents the activation energy? B
        (d)   What letter represents the heat of reaction? D

101

        (e)   Is the reaction endothermic or exothermic?
        (f)   What is the numerical value of ΔH?

8.   Draw the potential energy diagrams for two exothermic
      reactions having the same heats of reaction, but one of
      which is very fast at room temperature and one very slow.

9.3   Factors Affecting Reaction Rates

There are many factors that influence the rate at which a
chemical reaction will occur. The first is the nature of the
reactants themselves. This includes the stability of the
molecules, the type of bonding, the size and shape of the
molecules, and the physical states of the substances (whether
they are solids, liquids, or gases).

We have seen that for a reaction to occur the reactant particles
must collide. The second factor that affects the rate of a
reaction is the concentration of the reactants. The higher the
concentration of reactants, the larger the number of reactant
particles, and the higher the probability that a collision will
occur. The more collisions there are, the more likely there will
be a successful collision.

Only particles of a solid that are on the surface can undergo a
reaction. Increasing the surface area of a solid will increase
the number of particles on the surface. This increases the
number of particles available to react (the concentration of
reactant) and, therefore, increases the reaction rate.

Increasing the temperature of the reactants increases the kinetic
energy possessed by the reactant particles. This increases the
reaction rate for two reasons: it increases the number of
collisions and the energy of those collisions. In a similar
fashion, decreasing the temperature of a reaction will decrease
the rate of the reaction.

A catalyst is a substance that increases the rate of a chemical
reaction by lowering the activation energy barrier for the
reaction. Because the catalyst is not consumed in the reaction,
it can be used over and over again. This means it can be present
in very small amounts. Catalysts are widely used in the chemical
industry to increase yields and lower the cost of producing
chemicals. Most of the reactions that occur in our bodies would
proceed very slowly at body temperature without catalysts called
enzymes.

Important Terms

    catalyst   enzyme

9. What effect will each of the changes listed below have on the rate of the following reaction?

$$CaCO_3 + 2HCl \longrightarrow CaCl_2 + CO_2 + H_2$$
(Marble)

(a) Grind up the marble into fine particles. *Inc.*
(b) Use a lower concentration of hydrochloric acid. *decrease*
(c) Place the reaction container in an ice bath. *decrease*
(d) Add a catalyst. *Inc.*

10. When hydrogen iodide is heated to 400°C, it breaks apart to form hydrogen and iodine. Suppose you have a one liter flask containing one mole of HI at 400°C. What would be the effect of the following on the rate of the reaction?
(a) The volume of the flask is decreased to 500 mL while keeping the temperature constant. *Inc*
(b) You add another mole of HI to the flask. *Inc*
(c) The volume of the flask is increased to two liters without changing the temperature. *dec*
(d) The flask is warmed to 450°C. *Inc*
(e) The volume of the flask is increased to two liters and you add one mole of HI to the flask. The temperature remains 400°C. *dec* *No change*

11. Draw the potential energy diagram for an endothermic reaction. Using a dotted line, show what the potential energy diagram would look like if a catalyst were added to the reaction.

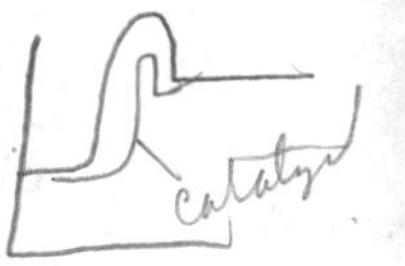

12. In what way can poisons affect the rate of a chemical reaction? *They destroy enzymes, decrease the rate of reactions in the body*

9.4 What is Chemical Equilibrium?

Chemical equilibrium is a dynamic state in which the rate of the forward reaction equals the rate of the reverse reaction. For a system in a chemical equilibrium, there is no net change in the number of reactant or product particles, even though the reactants are constantly forming products and the products are constantly forming reactants. For a reaction to reach equilibrium, two conditions must be met: a uniform temperature must be maintained and, after the reaction has started, no substances can be added or removed from the system.

When water is placed in an open container at 25°C, it will evaporate until all the water is gone. But if water is placed in a closed flask at 25°C from which all the gas has been evacuated, and if the pressure in the flask is then monitored, the pressure will first increase until it reaches 23.8 mm Hg and then will remain constant as long as the temperature remains constant. On the molecular level, the water molecules begin to evaporate as soon as the water is placed in the evacuated flask.

$$H_2O_{(l)} \longrightarrow H_2O_{(g)}$$

As the number of gas molecules increases, they begin to collide and condense to form liquid water.

$$H_2O_{(l)} \rightleftharpoons H_2O_{(g)}$$

The more gas molecules, the faster the rate of condensation until the rate of condensation equals the rate of evaporation. The system is then in equilibrium. This occurs when the pressure of the water vapor equals 23.8 mm Hg.

$$H_2O_{(l)} \rightleftharpoons H_2O_{(g)}$$

The number of water molecules entering the gas phase each second will equal the number of water molecules entering the liquid phase each second. Thus, equilibrium is a dynamic state in which changes occur at the same rate.

Important Terms

    equilibrium   dynamic state

Self-Test

13.   A chunk of iodine is placed in a water-alcohol mixture, and a reddish color quickly appears around the solid. The mixture is stirred and the color of the liquid initially deepens, and then no further change in the color of the liquid or in the mass of the iodine remaining on the bottom of the container can be detected. The temperature is held constant.
      (a)   Has an equilibrium been established in this situation? Give reasons for your answer. *Yes there is no change*
      (b)   State in words what has happened on the molecular level. *It reaches an equilibrium*
      (c)   State in equation form what has happened on the molecular level.

9.5   The Equilibrium Constant (optional section)

The value of the equilibrium constant, K_c, tells us how far toward completion a reaction will go before it reaches an

104

equilibrium. The larger the value of the equilibrium constant, the farther toward completion. The value of the equilibrium constant changes with changes in temperature. To determine the value of the equilibrium constant, divide the product of the molar concentration of the products raised to the power of the coefficients by the product of the molar concentration of the reactants raised to the power of the coefficients.

Example _____

Calculate the equilibrium constant for the following equilibrium, if at 25°C the molar concentration of dinitrogen tetroxide is 0.0277 M and of nitrogen dioxide is 0.0113 M.

$$N_2O_{4(g)} \rightleftharpoons 2NO_{2(g)} \qquad \text{at } 25°C$$

First we need to write the equilibrium constant expression for this reaction. Then, we substitute the values for the molar concentrations in the equation.

$$K_c = \frac{[NO_2]^2}{[N_2O_4]} = \frac{(0.0113)^2}{0.0277} = 4.61 \times 10^{-3}$$

9.6-9.7 Altering the Equilibrium: Le Chatelier's Principle

To alter or disrupt a chemical equilibrium, we might consider using the methods we discussed earlier for changing the rate of a chemical reaction: changing the concentration, changing the temperature, and adding a catalyst. The last method, adding a catalyst, will not be effective in disrupting an equilibrium because a catalyst increases the rate of both the forward and the reverse reaction by equal amounts. So we are left with changes in concentration and temperature as ways to disrupt an equilibrium. Le Chatelier's principle allows us to predict the effect of such changes on the equilibrium concentrations of reactants and products. Le Chatelier's principle states that if a stress is placed on a system in equilibrium, the system will change in a direction that will remove the stress.

Important Term

Le Chatelier's principle

Example _____

Suppose we have a flask containing dinitrogen tetroxide and nitrogen dioxide gases at equilibrium at 25°C. The equation for the reaction is

$$N_2O_{4(g)} \quad + \quad 13.9 \text{ kcal} \quad \rightleftharpoons \quad 2NO_2$$

colorless   reddish-brown

1. What effect would adding more N_2O_4 to the flask have on the equilibrium?

    Increasing the concentration of N_2O_4 will increase the rate of the forward reaction, forming more NO_2. With more NO_2 molecules, the rate of the reverse reaction will then increase until it again equals the rate of the forward reaction, and a new equilibrium will be established. The flask will be darker in color than before because it will contain more NO_2 molecules. The change in the system was the increase in concentration of N_2O_4. By Le Chatelier's principle, to resist the change, the system will shift to the right (the forward reaction will increase).

2. What would be the effect of removing some NO_2?

    Lowering the concentration of NO_2 will decrease the rate of the reverse reaction. As a result, the forward reaction will proceed at a faster rate until enough molecules of NO_2 have been formed to make the two rates again equal. The stress in this case was a decrease in the concentration of NO_2, and the system will shift to the right to produce more NO_2.

3. What would be the effect of doubling the pressure by decreasing the volume (with the temperature held constant)?

    The stress in this case is increased pressure. How can the system shift to remove the pressure? The forward reaction produces two molecules, while the reverse reaction produces only one. Since two molecules will exert twice as much pressure as one, the reverse reaction tends to form a system of lower pressure. Therefore, to remove the stress the system will shift to the left, producing N_2O_4 and causing the color to fade in the flask. A new equilibrium will be established with more N_2O_4 molecules and fewer NO_2 molecules.

4. What is the effect on the equilibrium of placing the flask in a water bath at 100°C?

    Increasing the temperature will increase the rate of both reactions, but the rate of the endothermic reaction (the reaction that removes heat) will be favored, forming more molecules of NO_2. The color of the flask will deepen until a new equilibrium is established. The stress on the system was the increased temperature, and the system will move in a direction that will remove the stress. The endothermic reaction removes energy and, therefore, is favored when the temperature is increased.

Can you predict what would happen to the color of the flask if it were placed in ice water?

Self-Test

Use the following equation to answer questions 14 to 17:

$$2NO \ + \ O_2 \ \rightleftharpoons \ 2NO_2 \ + \ 27 \ kcal$$

What would be the effect of each of the following on (a) the equilibrium concentration of O_2, and (b) the rate of formation of NO_2?

14. Increasing the concentration of NO (while the temperature remains constant).

15. Increasing the pressure (with constant temperature)

16. Increasing the temperature.

17. Adding a catalyst.

18. What is the effect of each of the changes below on the equilibrium concentration of HI in the following system?

$$H_2 \ + \ I_2 \ + \ 12.4 \ kcal \ \rightleftharpoons \ 2HI$$

(a) Increasing the pressure
(b) Removing some H_2
(c) Adding some I_2
(d) Increasing the temperature
(e) Adding a catalyst

Answers to Self-Test Questions in Chapter 9

1. F **2.** F **3.** T **4.** F **5.** T **6.** (a) exothermic (b) endothermic
(c) endothermic (d) exothermic **7.** (a) C (b) A (c) B (d) D
(e) endothermic (f) $\Delta H = (+)$ 16 kcal/mol
8.

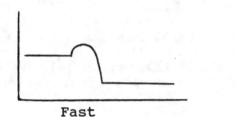

Fast   Slow

9. (a) increase (b) decrease (c) decrease (d) increase
10. (a) increase (b) increase (c) decrease (d) increase (e) no
change
11.

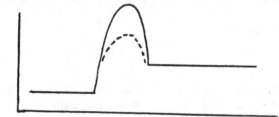

12. They destroy enzymes, thereby decreasing the rate of the
reactions in the body. **13.** (a) Yes. The temperature is constant
and no net change is observed in the system. (b) The solid
iodine began to dissolve. As the dissolved molecules began to
increase, some combined to reform the solid. A point was reached
when the rate of dissolving equaled the rate of reforming the
solid, and an equilibrium was established.
(c) Initially $I_{2(s)} \longrightarrow I_{2(aq)}$,   then $I_{2(s)} \rightleftharpoons I_{2(aq)}$.
    Finally $I_{2(s)} \rightleftharpoons I_{2(aq)}$
14. (a) decrease (b) increase **15.** (a) decrease (b) increase
16. (a) increase (b) increase **17.** (a) no change (b) increase
[Remember a catalyst will raise the rate of both the forward and
the reverse reactions] **18.** (a) no change [Note: Both sides
contain 2 molecules, so neither reaction will be favored]
(b) decrease (c) increase (d) increase (e) no change

Chapter 10 WATER, SOLUTIONS, AND COLLOIDS

Water is the dissolving fluid in all living organisms. An understanding of water and its properties, and of the properties of substances dissolved in water, is critical to our understanding of how living organisms function. This chapter discusses the properties of water, solutions, and colloids.

10.1 Molecular Shape of Water

Water has many properties that make it unique among chemicals, and essential to life processes. The water molecule is polar, and is an excellent solvent for ionic compounds and polar covalent compounds. Hydrogen bonds can form between water molecules or between water molecules and other molecules. It is the polarity of the water molecule and the possibility of hydrogen bonding that make water such an excellent solvent.

Important Terms

       solvent   universal solvent
       polar molecule   hydrogen bonding

10.2 Properties of Water

The polarity of the water molecule and the hydrogen bonds it can form are responsible for several unique properties of water. Water is an excellent solvent for ionic and polar compounds. The strong attraction between water molecules accounts for its high melting and boiling points when compared to other compounds with comparable molecular weights. Because of the hydrogen bonding between water molecules in ice, ice is less dense than water at 0°C. This is why ice floats on water. Water molecules at the surface are very strongly attracted to the other water molecules, and not to the nonpolar air molecules. This gives water a high surface tension. Water has a high heat of vaporization; that is, it takes a lot of energy to convert water from the liquid to the gaseous phase. Water also has a high heat of fusion; that is, water gives off a lot of heat when it goes from the liquid to the solid phase. Water has a high specific heat when compared to other liquids. This means that water can absorb heat without great changes in its temperature.

Important Terms

surface tension   heat of vaporization   heat of fusion
specific heat   surfactant

Self-Test

For questions 1 to 13, answer true (T) or false (F).

1. Humans can survive several weeks without water.

2. In the human body, water is found within the cells, around the cells, and in the blood plasma.

3. A water molecule is nonpolar, even though it contains two polar covalent bonds.

4. Hydrogen bonds can form between a hydrogen on one water molecule and a hydrogen on another water molecule.

5. Methane (CH_4) will dissolve in water.

6. The high melting point of water results from interactions between the water molecules.

7. The density of a substance is a measure of the mass of that substance per unit of volume.

8. Ice is more dense than water.

9. Because of its high surface tension, water will tend to spread out on a nonpolar surface and to form beads on a polar one.

10. Surfactants are substances that act to reduce the surface tension of water.

11. It takes 1.6 kilocalories to melt 20 grams of water.

12. A great deal of heat must be added to water to melt it or to vaporize it.

13. The higher the specific heat of a substance, the greater the change in its temperature when it absorbs a given amount of heat energy.

10.3 Suspensions

Three important mixtures are discussed in this chapter: suspensions, colloids, and solutions. The major difference between these three mixtures is the size of the dissolved particles (See Table 10.2).

Suspensions are heterogeneous mixtures whose particles will settle out in time and can be separated from the liquid with filter paper or a centrifuge.

Important Terms

suspension   heterogeneous

10.4 Colloids

The particles that form colloids (or colloidal dispersions) are larger than particles that form solutions. When placed in a solvent, they never truly dissolve; they are found suspended in the solvent, but are not heavy enough to settle out. Particles larger than colloids (those forming suspensions) will settle out of solution. Milk is a good example of a colloid. The white appearance of milk is caused by millions of tiny colloidal-size globules of fat suspended in the liquid. Table 10.3 in the text gives other examples of colloids.

Important Terms

colloidal dispersion   colloid

10.5 Properties of Colloids

Colloids exhibit some distinctive properties. Brownian movement is the random movement of colloidal particles caused by the bombardment of these particles by the solvent molecules. Colloids can be distinguished from true solutions by the Tyndall effect. In addition, the large surface area of colloidal particles gives them the ability to adsorb large amounts of other substances on their surfaces.

Important Terms

Brownian movement   Tyndall effect   adsorption

Self-Test _____

For questions 14 to 18, choose the correct answer.

14. Substances that form solutions when placed in water are
     (a) colloids   (c) colloidal dispersions
     (b) suspensions   (d) crystalloids

15. Heterogeneous mixtures containing particles suspended in a liquid that can be separated from the liquid using filter paper are called
    (a) colloids   (c) colloidal dispersions
    (b) suspensions   (d) crystalloids

16. If water contains a dispersion of particles larger than ions or molecules, but not large enough to settle out, the mixture is called a
    (a) colloid   (c) colloidal dispersion
    (b) suspension   (d) crystalloid

17. Which of the following is not a property of a colloid.
    (a) Brownian movement
    (b) adsorption
    (c) can be separated using filter paper
    (d) the Tyndall effect

18. You might use this to remove a colored impurity from a solution you were preparing in the laboratory.
    (a) filter paper   (c) centrifuge
    (b) powdered charcoal   (d) Brownian movement

10.6 Solutions

A solution is a homogeneous mixture whose particles are of atomic or molecular size. The solute is the substance that is being dissolved, and the solvent is the substance that does the dissolving. An aqueous solution is one in which water is the solvent.

Ionic solids will dissolve in water when the attraction between the ions and the water molecules is greater than the attraction among the ions in the crystal lattice. A hydrated ion is one that is surrounded by water molecules. Polar covalent substances are soluble in water because of polar-polar interactions and the hydrogen bonding that can form between these molecules and water molecules.

Important Terms

        solution   solute   solvent
        aqueous solution   hydrated ion

10.7 Electrolytes and Nonelectrolytes

Solutes, the particles dissolved in a solution, can be either non-electrolytes or electrolytes. Nonelectrolytes are uncharged solute particles whose aqueous solution will not conduct electricity. Sugar-water is a solution of a nonelectrolyte. Electrolytes are charged solute particles whose aqueous solutions will conduct electricity. Salt is an electrolyte. Electrolytes can occur in solution in two ways: through the breaking apart of an ionic crystal, or the ionization of a polar covalent compound. The charged solute particles formed will be anions (negative particles) and cations (positive particles). A strong electrolyte is one that completely ionizes or breaks apart in solution; a weak electrolyte only partially ionizes or breaks apart.

Important Terms

electrolyte   nonelectrolyte   cation   anion

Self-Test _____

19. Predict whether each of the following are electrolytes (E) or nonelectrolytes (N):

    (a) KCl   (b) $CaSO_4$ (c) HBr   (d) CH_3CH_2OH (alcohol)

20. Indicate which of the following are cations and which are anions.

    (a) Li (b) SO_4^{2-} (c) I^- (d) NH_4^+

10.8 Factors Affecting the Solubility of a Solute

The solubility of a solute indicates the amount of solute that will dissolve in a solvent. Several factors affect the solubility of a solute: the nature of the solute, the temperature, and the pressure (if the solute is a gas.)

Important Term

solubility

10.9 The Solubility of Ionic Solids

Ionic solids vary in their solubility. The general rules found in Table 10.4 allow us to predict whether a precipitate will form when two solutions of soluble ionic solids are mixed.

Important Terms

net-ionic equation   precipitate

Will a precipitate form if a solution of lead nitrate and sodium sulfide are mixed?

A solution of lead nitrate, $Pb(NO_3)_2$, contains lead ions (Pb^{2+}) and nitrate ions (NO_3^-). A solution of sodium sulfide, Na_2S, contains sodium ions (Na^+) and sulfide ions (S^{2-}). The two new substances that could form from a mixture of these two solutions are $NaNO_3$ and PbS. Looking at Table 10.4 in the text, we see that $NaNO_3$ is soluble but PbS is not and will form a precipitate. The complete equation for this reaction would be

$$Pb^{2+}_{(aq)} + 2NO_3^-{}_{(aq)} + 2Na^+_{(aq)} + S^{2-}_{(aq)} \longrightarrow PbS_{(s)} + 2Na^+_{(aq)} + 2NO_3^-{}_{(aq)}$$

To obtain the net-ionic equation, you cancel the ions not involved in the reaction.

$$Pb^{2+}_{(aq)} + S^{2-}_{(aq)} \longrightarrow PbS_{(s)}$$

Self-Test

For questions 21 to 24, (a) predict whether a precipitate will form when aqueous solutions containing the two compounds are mixed, and (b) write the net-ionic equation when a reaction does occur.

21. Potassium sulfide and silver nitrate

22. Ammonium chloride and sodium acetate

23. Zinc bromide and sodium carbonate

24. Barium nitrate and lithium hydroxide

Answers to Self-Test Questions in Chapter 10

1. F 2. T 3. F 4. F 5. F 6. T 7. T 8. F 9. F 10. T
11. T 12. T 13. F 14. d 15. b 16. a 17. c 18. b 19. (a) E
(b) E (c) E (d) N 20. (a) cation (b) anion (c) anion (d) cation
21. (a) Yes (b) $S^{2-}_{(aq)} + 2Ag^+_{(aq)} \longrightarrow Ag_2S_{(s)}$
22. (a) No 23. (a) Yes (b) $Zn^{2+}_{(aq)} + CO_3^{2-}{}_{(aq)} \longrightarrow ZnCO_{3(s)}$
24. (a) Yes (b) $Ba^{2+}_{(aq)} + 2OH^-_{(aq)} \longrightarrow Ba(OH)_{2(s)}$

Chapter 11   SOLUTION CONCENTRATIONS

The concentration of solute particles in a solution is often critical to the normal functioning of living organisms. In this chapter we study several methods used to measure the concentration of a solute in a solution.

11.1 Saturated and Unsaturated Solutions

The concentration of a solution indicates the number of solute particles dissolved in the solvent. If the solution is dilute, there are few solute particles; if it is concentrated, there are many. If it is saturated, the solvent contains all the solute particles it can hold at that temperature. A supersaturated solution is an unstable solution that results when the temperature of a saturated solution is lowered but no crystals form. Crystals will form very rapidly when a small crystal is added.

Important Terms

       dilute   concentrated   saturated
       supersaturated   unsaturated

11.2 Molar Concentration

The concentration of a solution is a measure of the relative amount of solute in the solution, and is always expressed as a ratio. The molar concentration or molarity of a solution is defined as the number of moles of solute per liter of solution. The concentration of the majority of solutions you will use in the laboratory are expressed in molarity.

Important Term

       molarity

Example _____

1.   What does "0.5 M $MgCrO_4$" on a bottle label tell you about the solution in that bottle?

      The solution will have 0.5 mole, or 70 grams, of $MgCrO_4$ dissolved in each liter of solution.

2.   How would you make up 200 mL of 0.50 M KCl?

0.50 M means 0.50 mole per liter of solution. To make one liter of solution you would need 0.50 mole of KCl.

$$0.50 \text{ mole KCl} \times \frac{74.6 \text{ g}}{1 \text{ mol KCl}} = 37.3 \text{ g KCl}$$

But the problem specifies only 200 mL of solution.

$$\frac{37.3 \text{ g}}{1000 \text{ mL}} \times 200 \text{ mL} = 7.5 \text{ g}$$

So, to make 200 mL of 0.50 M KCl you would add 7.5 grams of KCl to enough water to make 200 mL of solution.

3.   What is the molarity of 500 mL of solution that contains 1.7 grams of ammonia?

The first step is to determine the number of moles of ammonia in 1.7 grams.

$$1.7 \text{ g} \times \frac{1 \text{ mole NH}_3}{17 \text{ g}} = 0.10 \text{ mol of NH}_3$$

Our solution contains 0.10 mole in 500 mL. Therefore, one liter or 1000 mL would contain 0.20 mole. Thus the concentration is 0.20 M NH_3.

Self-Test _____

For questions 1 to 4, calculate the molarity of the solution.

1.   9.0 grams of $C_6H_{12}O_6$ in 500 mL of solution.

2.   120 grams of NaOH in 2.0 liters of solution.

3.   0.14 mole of HCl/100 mL of solution.

4.   87.7 grams of NaCl in 750 mL of solution.

For questions 5 to 8, indicate how you would prepare each solution.

5.   0.800 liter of 2.45 M H_3PO_4

6.   375 mL of 0.250 M KOH

7.   100 mL of 6.00 M H_2SO_4

8.   0.25 liter of 0.10 M NaOH

9. How many milliliters of 6.0 M HCl contain 0.40 mole of HCl?

11.3 Percent Concentration

Percent concentrations are units of concentration that do not take into account the formula weight of the solute. Two types of percent concentration are discussed in this section:

Weight/volume (w/v) percent is defined as the number of grams of solute per 100 mL of solution.

Milligram (mg%) percent is defined as the number of milligrams of solute per 100 mL of solution.

Important Terms

weight/volume percent   milligram percent

Self-Test

For questions 10 to 13, describe how you would prepare each of the solutions.

10. 225 mL of 2.0% (w/v) glucose $225 \, ML \times \dfrac{2.0 \, g}{1000 \, ML} = 4.5$

11. 0.030 liter of 0.70% (w/v) $MgSO_4$

12. 5 deciliters of 8.0 mg% sucrose

13. 0.040 liter of 34 mg% $CaCl_2$

14. What is the concentration in w/v percent of 500 mL of solution that contains 920 mg of $C_6H_{12}O_6$?

15. What is the concentration of Na^+ in mg% if 5.0 mL of blood contains 0.14 mg of Na^+?

16. What is the concentration of NaCl in w/v percent if 750 mL of solution contain 1.50 mole of NaCl?

For questions 17 to 25, fill in the blank with the correct word or words.

17. A *dilute* solution contains very few solute particles, and a *concentrated* solution contains many solute particles.

18. In a *saturated* solution the solute particles in solution are in equilibrium with the undissolved solute particles in the container.

117

19. A supersaturated solution can form when the temperature of a saturated solution is _lowered_ and no crystals form.

20. The unit of concentration that expresses concentration in moles per liter of solution is _molarity_

21. _W V %_ gives the concentration of a solution in grams of solute per 100 mL of solution.

22. Concentration in _molarity_ takes into account the formula weight of the solute.

23. _percent_ concentrations do not take into account the formula weight of the solute.

24. The concentrations of trace minerals in the blood are often expressed in units of _Mg %_.

25. Clinical reports often have concentrations expressed in _W/V Mg%_.

11.4 Parts per Million and Parts per Billion

Parts per million (ppm) and parts per billion (ppb) are units of concentration used to describe extremely dilute solutions. You will find these terms used quite often to describe pollutants in water and air.

Important Terms

parts per million (ppm)   parts per billion (ppb)

Example _____

A sample of water was found to contain carbon tetrachloride in a concentration of 2 ppb. How many micrograms of carbon tetrachloride would be found in a 100 mL-sample?

$$1 \text{ ppb} = \frac{1 \ \mu g}{1000 \text{ mL}}, \quad \text{therefore } 2 \text{ ppb} = \frac{2 \ \mu g}{1000 \text{ mL}}$$

A 100 mL sample would contain $\frac{2 \ \mu g}{1000 \text{ mL}} \times 100 \text{ mL} = 0.2 \ \mu g$

Self-Test _____

26. A sample of waste water from a film processing plant contained silver in a concentration of 4 ppm. How many milligrams of silver would be found in a 200-mL sample?

118

27. A 500-mL sample from a river in Japan was found to have 0.02 mg of cadmium. What is the concentration of cadmium in both parts per million and parts per billion?

11.5 Equivalents

The ionic components of a solution are often described in terms of equivalents. One equivalent is equal to 1 mole of charge. The gram-equivalent weight of a substance is equal to the number of grams of that substance that will contain one equivalent, or one mole of charge. The ionic components of the blood are reported in milliequivalents (1000 mEq = 1 Eq).

Important Terms

equivalent (Eq)   gram-equivalent weight
milliequivalent (mEq)

Example _____

A 100-mL sample of blood was found to contain 0.14 mEq of sodium ions. How many milligrams of sodium were there in the blood sample?

The sodium ion is Na^+. 1 mol Na^+ contains 1 mol of + = 1 Eq

The gram-equivalent weight of Na^+ = the weight of 1 mole
= 23 g.

1 mEq of Na^+ will contain $\dfrac{23 \text{ g}}{1 \text{ Eq}} \times \dfrac{1 \text{ Eq}}{1000 \text{ mEq}} \times \dfrac{1000 \text{ mg}}{1 \text{ g}} = \dfrac{23 \text{ mg}}{1 \text{ mEq}}$

The blood sample contained $0.14 \text{ mEq} \times \dfrac{23 \text{ mg}}{1 \text{ mEq}} = 3.2 \text{ mg}$

Self-Test _____

28. How many milligrams of potassium ions are there in a sample of blood that contains 2.5 mEq of potassium ions?

29. How many milliequivalents of magnesium ions are present in 75 mL of 0.80% (w/v) Mg^{2+} solution?

11.6 Dilutions

You will often be required to make up a solution by diluting a more concentrated stock solution.

How would you prepare 70 mL of 2% Na_2CO_3 from a 5% Na_2CO_3 stock solution?

To solve this problem we substitute the given values in equation 1 from Section 11.6 of the text.

$$2\% \ Na_2CO_3 \times 70 \ mL = 5\% \ Na_2CO_3 \times V$$

$$\frac{2\% \ Na_2CO_3 \times 70 \ mL}{5\% \ Na_2CO_3} = V$$

$$28 \ mL = V$$

To make the desired solution you would measure out 28 mL of the stock solution, then add enough water to make 70 mL of solution.

Self-Test

30. How many milliliters of a 6.0 M HBr stock solution would you need to prepare 15 mL of 2.5 M HBr?

31. How many milliliters of a 10% NaCl solution would you need to prepare 0.50 liters of physiological saline (0.90% NaCl)?

32. What is the final concentration of a solution made from a 1:4 dilution of 3.6 M NaOH?

11.7 Colligative Properties

Properties of solutions that depend only upon the number of solute particles in the solution are called colligative properties. Such colligative properties include the raising of the boiling point and the lowering of the freezing point of water by solute particles.

Important Term

colligative property

11.8 Osmosis

Another colligative property is osmotic pressure. Osmosis is the migration of water molecules through a differentially permeable membrane (one that will let water, but not solute particles, pass through) from a region of lower solute concentration to a region

of higher solute concentration. Osmotic pressure is defined as
the amount of pressure that must be applied to prevent this flow
of water. The higher the solute concentration, the greater the
osmotic pressure.

Important Terms

osmosis   osmotic pressure

11.9 Osmolarity

The osmolarity of a solution indicates the total number of moles
of all solute particles in one liter of solution. One osmol is
one mole of any combination of solute particles (molecules or
ions). Solutions having different types of solute particles but
the same osmolarity will have the same osmotic pressure. The
higher the osmolarity of a solution, the greater the osmotic
pressure of that solution.

Important Terms

osmolarity   osmol   milliosmol (mOsm)

Example ──

What is the osmolarity of one liter of solution containing one
mole of calcium nitrate and one mole of glucose?

> 1 mole of $Ca(NO_3)_2$ will dissolve in solution to form 1 mole
> of Ca^{2+} ions and 2 moles of NO_3^- ions. The glucose molecule
> doesn't ionize when dissolved in water, so there will be 1
> mole of glucose molecules in solution. The total number of
> moles of particles in solution is four. Therefore, the
> osmolarity of the solution is 4 osmols/liter.

11.10 Isotonic Solutions

The movement of water is critical to living organisms. If the
concentration of the solution outside a cell is equal to the
concentration inside the cell, the solution is said to be
isotonic. In an isotonic solution, the water will move in and
out of the cell at the same rate. If the concentration of the
solution outside the cell is greater than that inside the cell,
the solution is hypertonic to the cell; water will flow out of
the cell, causing the cell to shrink. If the concentration of
the solution outside the cell is less than that inside the cell,
the solution is hypotonic to the cell; water will move into the
cell, causing the cell to swell. Electrolytes in body fluids are
important in maintaining the osmotic balance in the body.

Important Terms

> isotonic   hypertonic   hypotonic   edema
> hemolysis   crenation   normal saline

11.11   Dialysis

Dialysis is the movement of ions and small molecules through membranes called dialyzing membranes. Colloids can be separated from crystalloids using dialysis. Hemodialysis is a procedure that makes use of dialysis to remove waste products from the blood of patients suffering from kidney failure.

Important Terms

> dialysis   hemodialysis   dialyzing membrane

Self-Test

For questions 33 to 40, answer true (T) or false (F).

33. Osmosis is the movement of water through a membrane from a region of low solute concentration to a region of higher solute concentration.

34. Solutions whose osmotic pressures are equal are said to be isotonic.

35. Red blood cells placed in a 0.2 M NaCl solution will undergo crenation.

36. Red blood cells placed in a 200 mL solution containing 40 grams of NaCl will undergo hemolysis.

37. No change will be observed in red blood cells placed in 0.5 liter of solution containing 4.5 grams of NaCl.

38. A solution of 2 M K_3PO_4 will have an osmolarity of 4 osmols/liter.

39. 100 mL of 3 M NaCl will have the same osmotic pressure as 100 mL of 3 M NaOH.

40. If a dialysis bag containing an aqueous mixture of sodium chloride, glucose ($C_6H_{12}O_6$), and protein were placed in a beaker of water, the water in the beaker would eventually contain sodium chloride and glucose but not protein.

Answers to Self-Test Questions in Chapter 11

1. $9.0 \text{ g} \times \dfrac{1 \text{ mol}}{180 \text{ g}} = 0.050 \text{ mol}$

   $\dfrac{0.050 \text{ mol}}{500 \text{ mL}} \times \dfrac{1000 \text{ mL}}{1 \text{ liter}} = \dfrac{0.10 \text{ mol}}{1 \text{ liter}} = 0.10 \text{ M } C_6H_{12}O_6$

2. $120 \text{ g} \times \dfrac{1 \text{ mol}}{40 \text{ g}} = 3.0 \text{ mol}; \quad \dfrac{3.0 \text{ mol}}{2.0 \text{ liter}} = \dfrac{1.5 \text{ mol}}{1 \text{ liter}} = 1.5 \text{ M NaOH}$

3. $\dfrac{0.14 \text{ mol}}{100 \text{ mL}} \times \dfrac{1000 \text{ mL}}{1 \text{ liter}} = \dfrac{1.4 \text{ mol}}{1 \text{ liter}} = 1.4 \text{ M HCl}$

4. $87.7 \text{ g} \times \dfrac{1 \text{ mol}}{58.5 \text{ g}} = 1.50 \text{ mol}$

   $\dfrac{1.50 \text{ mol}}{750 \text{ mL}} \times \dfrac{1000 \text{ mL}}{1 \text{ liter}} = \dfrac{2.00 \text{ mol}}{1 \text{ liter}} = 2.00 \text{ M NaCl}$

5. $0.800 \text{ liter} \times \dfrac{2.45 \text{ mol } H_3PO_4}{1 \text{ liter}} \times \dfrac{98.0 \text{ g}}{1 \text{ mol } H_3PO_4} = 192 \text{ g}$

   Add 192 g of H_3PO_4 to enough water to make 0.800 liter of solution.

6. $375 \text{ mL} \times \dfrac{0.250 \text{ mol KOH}}{1000 \text{ mL}} \times \dfrac{56.1 \text{ g}}{1 \text{ mol KOH}} = 5.26 \text{ g}$

   Add 5.26 g of KOH to enough water to make 375 mL of solution.

7. $100 \text{ mL} \times \dfrac{6.00 \text{ mol } H_2SO_4}{1000 \text{ mL}} \times \dfrac{98.1 \text{ g}}{1 \text{ mol } H_2SO_4} = 58.9 \text{ g}$

   Add 58.9 g of H_2SO_4 to enough water to make 100 mL of solution.

8. $0.25 \text{ liter} \times \dfrac{0.10 \text{ mol NaOH}}{1 \text{ liter}} \times \dfrac{40 \text{ g}}{1 \text{ mol NaOH}} = 1.0 \text{ g}$

   Add 1.0 g NaOH to enough water to make 0.25 liter of solution.

9. $0.40 \text{ mol} \times \dfrac{1 \text{ liter}}{6.0 \text{ mol}} \times \dfrac{1000 \text{ mL}}{1 \text{ liter}} = 67 \text{ ml}$

10.
$$225 \text{ mL} \times \frac{2.0 \text{ g glucose}}{1000 \text{ mL}} = 4.5 \text{ g glucose}$$

Add 4.5 g glucose to enough water to make 225 mL solution.

11.
$$0.030 \text{ liter} \times \frac{1000 \text{ mL}}{1 \text{ liter}} \times \frac{0.70 \text{ g MgSO}_4}{100 \text{ mL}} = 0.21 \text{ g MgSO}_4$$

Add 0.21 g MgSO$_4$ to enough water to make 0.030 liter of solution.

12.
$$5 \text{ dL} \times \frac{100 \text{ mL}}{1 \text{ dL}} \times \frac{8 \text{ mg sucrose}}{100 \text{ mL}} = 40 \text{ mg sucrose}$$

Add 40 mg of sucrose to enough water to make 5 dL of solution.

13.
$$0.040 \text{ liter} \times \frac{1000 \text{ mL}}{1 \text{ liter}} \times \frac{34 \text{ mg CaCl}_2}{100 \text{ mL}} = 14 \text{ mg CaCl}_2$$

Add 14 mg CaCl$_2$ to enough water to make 0.040 liter of solution.

14.
$$920 \text{ mg} \times \frac{1 \text{ g}}{1000 \text{ mg}} \ ; \quad \frac{0.920 \text{ g C}_6\text{H}_{12}\text{O}_6}{500 \text{ mL}} = \frac{X}{100 \text{ mL}}$$

$$X = 0.184 \text{ g}$$

$$\frac{0.184 \text{ g}}{100 \text{ mL}} = 0.184\% \text{ C}_6\text{H}_{12}\text{O}_6$$

15.
$$\frac{0.14 \text{ mg Na}^+}{5.0 \text{ mL}} = \frac{X}{100 \text{ mL}} \ ; \ X = 2.8 \text{ mg Na}^+ \ ; \ 2.8 \text{ mg\%}$$

16.
$$1.50 \text{ mol} \times \frac{58.5 \text{ g NaCl}}{1 \text{ mol}} = 87.8 \text{ g}; \quad \frac{87.8 \text{ g NaCl}}{750 \text{ mL}} = \frac{X}{100 \text{ mL}}$$

$$X = 11.7\% \text{ NaCl}$$

17. dilute, concentrated **18.** saturated **19.** lowered
20. molarity **21.** weight/volume percent **22.** molarity
23. percent **24.** milligram percent **25.** weight/volume or milligram percent

26.
$$200 \text{ mL} \times \frac{4 \text{ mg Ag}}{1 \text{ liter}} \times \frac{1 \text{ liter}}{1000 \text{ mL}} = 0.8 \text{ mg Ag}$$

27.
$$\frac{0.02 \text{ mg Cd}}{500 \text{ mL}} \times \frac{1000 \text{ mL}}{1 \text{ liter}} = \frac{0.04 \text{ mg Cd}}{1 \text{ liter}} = 0.04 \text{ ppm Cd}$$

$$\frac{0.04 \text{ mg Cd}}{1 \text{ liter}} \times \frac{1000 \ \mu g}{1 \text{ mg}} = \frac{40 \ \mu g \text{ Cd}}{1 \text{ liter}} = 40 \text{ ppb Cd}$$

28. $2.5 \text{ mEq K}^+ \times \dfrac{39.1 \text{ g}}{1 \text{ Eq}} \times \dfrac{1 \text{ Eq}}{1000 \text{ mEq}} \times \dfrac{1000 \text{ mg}}{1 \text{ g}} = 98 \text{ mg K}^+$

29. $0.80\% \text{ Mg}^{2+} = \dfrac{0.80 \text{ g Mg}^{2+}}{100 \text{ mL}} \times 75 \text{ mL} = 0.60 \text{ g Mg}^{2+}$

$0.60 \text{ g Mg}^{2+} \times \dfrac{1 \text{ Eq}}{12.2 \text{ g}} \times \dfrac{1000 \text{ mEq}}{1 \text{ Eq}} = 49 \text{ mEq Mg}^{2+}$

30. $2.5 \text{ M} \times 15 \text{ mL} = 6.0 \text{ M} \times \text{V}; \quad \text{V} = 6.2 \text{ mL}$
31. $0.50 \text{ liter} \times 0.90\% \text{ NaCl} = 10\% \text{ NaCl} \times \text{V}$
    $\text{V} = 0.045 \text{ liter} = 45 \text{ mL}$
32. $4\text{V} \times \text{C} = 1\text{V} \times 3.6 \text{ M NaOH}; \quad \text{C} = 0.90 \text{ M NaOH}$
33. T 34. T 35. T 36. F 37. T 38. F 39. T 40. T

Chapter 12   ACIDS AND BASES

In this chapter we study two very important classes of compounds: acids and bases. We discuss their properties, their interactions, the methods used to describe and measure their concentrations, and the way in which living organisms protect themselves against large changes in the concentrations of acid and base.

12.1   Acids and Bases: the Brønsted-Lowry Definition

Acids and bases play an important role in body chemistry. Acids form aqueous solutions that taste sour, turn litmus paper red, and react with metals to produce hydrogen. Bases form aqueous solutions that taste bitter, feel slippery to the touch, and turn litmus paper blue.

There are several definitions of acids and bases, but we will define an acid as a substance that donates protons (the positive hydrogen ion, H^+), and a base as a substance that accepts protons. This definition is the Brønsted-Lowry definition of acids and bases. In this definition we talk about conjugate acid-base pairs. The conjugate base of an acid is the negative ion that results when the acid donates a hydrogen ion. Consider the following equation:

$$H_2CO_3 \; + \; H_2O \; \rightleftharpoons \; H_3O^+ \; + \; HCO_3^-$$

$$\text{acid}_1 \qquad \text{base}_2 \qquad \text{acid}_2 \qquad \text{base}_1$$

The bicarbonate ion (HCO_3^-) is the conjugate base of carbonic acid (H_2CO_3.) The hydronium ion is the conjugate acid of water, which is acting as a base in this reaction.

Important Terms

acid	base	conjugate acid-base pair
hydronium ion	hydroxide ion	polyprotic acid
amphoteric		

Self-Test _____

1. List three characteristics of (a) acidic solutions and (b) basic solutions.

For questions 2 to 5, identify the two conjugate acid-base pairs:

2.   HF   + NH_3 $\rightleftharpoons$ NH_4^+ + F^-
     A1   A2   B2   B1

3. N_2H_4 + H_2O $\rightleftharpoons$ $N_2H_5^+$ + OH^-
     B1   A2   A1   B2

4. H_2SO_4 + H_2O $\rightleftharpoons$ H_3O^+ + HSO_4^-

5. H_2SO_3 + HNO_3 $\rightleftharpoons$ $H_3SO_3^+$ + HNO_3^-

12.2 Strength of Acids and Bases

A strong acid is one that readily donates its protons.
Therefore, its conjugate base will have a weak attraction for
protons: strong acid—weak conjugate base. A weak acid does not
donate its protons as readily. As a result, its conjugate base
has a strong attraction for protons: weak acid—strong conjugate
base. The same is true for bases. A strong base has a large
attraction for protons and its conjugate acid will be weak. A
weak base will not attract protons as readily, and its conjugate
acid will be strong.

Self-Test

6.   Identify the following as an acid or a base, then list the
     acids and the bases in order of increasing strength:

     OH^-, NO_3^-, H2O, H_2SO_4, HPO_4^{2-}, PO_4^{3-}, HCl, HNO_3, H_2CO_3, CO_3^{2-}

12.3 Naming Acids

Binary acids (acids containing hydrogen and one other nonmetallic
element) are named by adding the prefix, hydro-, to the name of
the second element. The name of the second element is given an
-ic ending and then the word "acid" is added. The name of salts
formed from binary acids end in -ide. For example, HCl is
hydrochloric acid and its potassium salt is NaCl, sodium
chloride.

Acids containing hydrogen, oxygen, and a third nonmetallic
element (called oxoacids) are named by adding the suffix -ic or
-ous to the name of the third element and then adding the name
"acid". The -ic ending is used for the acid containing the third
nonmetallic element with the higher oxidation number. The name
of polyatomic ions formed from oxoacids ending in "-ic" end in
"-ate". Those salts formed from oxoacids ending in "-ous" end in
"ite". For example, consider the two oxoacid $HClO_2$ and $HClO_3$.
The chlorine in $HClO_2$ has an oxidation number of 3+ and in $HClO_3$

127

an oxidation number of 5+. The name for $HClO_2$ is chlorous acid and $HClO_3$ is chloric acid. The sodium salt of chlorous acid is sodium chlorite, $NaClO_2$, and the sodium salt of chloric acid is sodium chlorate, $NaHClO_3$.

Important Terms

binary acids   oxoacids   acid salts

Example

1. What is the name of the compound HBr?

    This is a binary acid, so the name is be hydrobromic acid.

2. Name the following acids: H_2SO_3 and H_2SO_4.

    These acids contain the same three elements. The oxidation number of sulfur in the first acid is 4+ and in the second acid is 6+. Therefore, the first acid will be named with an −ous ending and the second with an −ic ending.

    H_2SO_3 is sulfurous acid H_2SO_4 is sulfuric acid

Self-Test

7. Name the following compounds:
    (a) HF   (c) KH_2PO_4 (e) HNO_2
    (b) HNO_3 (d) $NaHSO_3$ (f) $CaSO_4$

12.4 Neutralization Reactions

An acid will react with a base to produce water and a salt. A neutral solution is one which contains equal amounts of hydrogen ions and hydroxide ions. As a result, it will have neither acidic nor basic properties. A neutralization reaction is one in which a solution of acid reacts with a solution of base to produce a neutral solution.

Important Terms

neutral solution   neutralization   salt

Self-Test

8. Write the balanced equation and the net-ionic equation for:
    (a) The neutralization of $Ca(OH)_2$ by HCl.
    (b) The neutralization reaction between HNO_3 and KOH

12.5 Ionization of Water

Water molecules can be ionized by other water molecules to form the hydrogen ion and the hydroxide ion. Although we use the term hydrogen ion or proton when talking about acids and bases, the hydrogen ion does not exist as a separate particle in aqueous solution. It is always joined to a water molecule in the form of the hydronium ion, H_3O^+. The hydronium ion contains a coordinate covalent bond in which the oxygen donates both electrons found in the bond between the hydrogen and the oxygen.

$$H_2O \ + \ H_2O \ \rightleftharpoons \ H_3O^+ \ + \ OH^-$$

$$\text{Hydronium ion} \qquad \text{Hydroxide ion}$$

Important Terms

hydrogen ion
coordinate covalent bond

hydroxide ion
hydronium ion

12.6 Ion Product of Water, K_w

In pure water the concentration of the hydrogen ion times the concentration of the hydroxide ion equals a constant called the ion product of water, K_w. The numerical value of K_w is 1×10^{-14}. Therefore,

$$[H^+][OH^-] = 1 \times 10^{-14}$$

Important Term

ion product of water, K_w

12.7 The pH Scale

The molarity of solutions of acids or bases found in living organisms normally falls within certain ranges. A scale called the pH scale was devised as an easy way to indicate the concentration of the hydrogen ion in solution. The pH scale most commonly used ranges from a concentration of 1 M H^+ to 1×10^{-14} M H^+ (pH = 0 to pH = 14). Pure water, which is neutral, will have a pH of 7. Study carefully Table 12.2 and Figure 12.3 in the text for the conversions between concentration and pH.

The pH of a solution can be measured in many ways, from fairly qualitative methods using indicator dyes to the very accurate measurements of a pH meter.

Important Terms

pH   pH meter   acid-base indicator

12.8 Titrations

Normality is a unit of concentration that indicates the number of
hydrogen or hydroxide ions present in a solution. Normality is
the number of equivalents of acid or base per liter of solution.
An equivalent of acid is the amount of acid that will donate one
mole of hydrogen ions. An equivalent of base is the amount of
base that will neutralize one mole of hydrogen ions.

The unknown concentration of an acid or base solution can be
determined using a procedure called titration. In a titration,
an acidic or basic solution of known concentration is added to
the unknown solution until equal amounts of acid and base are in
the container. That point, called the equivalence point, is
determined by an acid-base indicator or a pH meter. From the
number of milliliters of known solution added, and from the known
concentration of this acid or base, the concentration of the
unknown base or acid can be determined.

$$N_a \times V_a = N_b \times V_b$$

Important Terms

normality (N)   equivalent (Eq)   titration
equivalence point   end point

Example _____

1.   100 mL of urine from a patient with kidney trouble is
     titrated with 0.002 N base. The indicator changes color
     when 5.0 mL of base has been added. What is the acid
     concentration and the pH of the patient's urine?

     0.002 N base contains 0.002 equivalents of OH⁻ in 1000 mL.
     Therefore, 5 mL will contain

$$5 \text{ mL} \times \frac{0.002 \text{ Eq}}{1000 \text{ mL}} = 0.00001 \text{ Eq} = 1 \times 10^{-5} \text{ Eq OH}^-$$

     At the equivalence point of a titration, the equivalents of
     base equal the equivalents of acid. One equivalent of acid
     is define as the amount of acid that will donate one mole of
     H^+. So the urine contains 1×10^{-5} moles H^+ in 100 mL.

$$[H^+] = \frac{1 \times 10^{-5}}{100 \text{ mL}} = \frac{1 \times 10^{-4}}{1000 \text{ mL}} ; \quad pH = 4$$

2. What is the weight of one equivalent of sulfuric acid, H_2SO_4?

One mole of H_2SO_4 will donate two moles of hydrogen ions, so we know

$$1 \text{ mol } H_2SO_4 = 2 \text{ Eq } H_2SO_4$$

and

$$1 \text{ mol } H_2SO_4 = 98.1 \text{ g}$$

Therefore, the weight of one equivalent of H_2SO_4 is

$$\frac{98.1 \text{ g}}{1 \text{ mol } H_2SO_4} \times \frac{1 \text{ mol } H_2SO_4}{2 \text{ Eq } H_2SO_4} = \frac{49.0 \text{ g}}{1 \text{ Eq } H_2SO_4}$$

3. What is the normality of a solution containing 24.3 g H_2SO_4 in 150 mL of solution?

We need to determine the number of equivalents of H_2SO_4 in one liter of this solution.

$$\frac{24.3 \text{ g}}{150 \text{ mL}} \times \frac{1 \text{ Eq } H_2SO_4}{49.0 \text{ g}} \times \frac{1000 \text{ mL}}{1 \text{ liter}} = \frac{3.31 \text{ Eq } H_2SO_4}{1 \text{ liter}} = 3.31 \text{ N } H_2SO_4$$

4. What is the normality of base if 100 mL of the solution of base is completely neutralized by 58 mL of 2.4 N H_2SO_4?

Substituting in the equation for titrations, we have

$$2.4 \text{ N} \times 58 \text{ ml} = N_b \times 100 \text{ mL}$$

$$1.4 \text{ N} = N_b$$

Self-Test ———————————————————————————————

For questions 9 to 16, give the (a) $[H^+]$, (b) $[OH^-]$, (c) pH, and (d) indicate if the solution is acidic or basic.

9. 0.1 M HCl

10. 1×10^{-4} M HNO_3

11. 0.001 M NaOH

12. 1×10^{-6} M KOH

13 0.0001 M lithium hydroxide

131

14. 36.5 μg HCl in 100 mL of solution

15. 370 mg of Ca(OH)$_2$ in one liter of solution. (Hint: Each mole of Ca(OH)$_2$ yields 2 moles of OH$^-$ in solution.)

16. A solution made by adding 50 mL of 0.02 M NaOH to 50 mL of 0.04 M HCl.

17. What is the concentration of acid (HCl) in 100 mL of stomach fluid if neutralization occurs when 15 mL of 0.20 M NaOH is added?

18. Give the weight of one equivalent for the following acids or bases:
    (a) HBr   (b) NaOH   (c) H$_2$SO$_3$ (d) Ca(OH)$_2$

19. How would you prepare 300 mL of 0.150 N HBr?

20. What is the normality of a solution that contains 0.18 g Ca(OH)$_2$ in 100 mL of solution?

21. What is the normality of a sulfurous acid solution if 50 mL of this solution is completely neutralized by 37 mL of 0.15 N NaOH?

12.9 What Are Buffers?

Buffers are substances that, when placed in solution, protect against changes in pH. The best buffer systems consist of a weak acid and its conjugate base, or a weak base and its conjugate acid. These systems have their greatest buffering capacity at a pH where the concentration of the acid equals the concentration of its conjugate base.

Important Term

buffer

12.10   Control of pH in Body Fluids

Buffers in the blood plasma help keep the pH of the blood in a very narrow range even if there is an increase in acid or base in the blood. Reread carefully the discussion in the text of the carbonic acid-bicarbonate buffer system.

Acidosis is a condition that results when the pH of the blood is lower than normal (more acidic than normal). The cause of acidosis can be respiratory or metabolic. When the pH of the blood is higher (more basic) than normal, a more rare condition called alkalosis results.

Important Terms

acidosis alkalosis

Self-Test

For questions 22 to 26, choose the best answer or answers from the choices given.

22. Substances that protect the blood from large changes in pH are called
    (a) electrolytes   (c) acids
    (b) bases   (d) buffers

23. A drop in the alkaline reserves of the blood will cause a condition called
    (a) acidosis   (c) alkalosis
    (b) edema   (d) emphysema

24. Hyperventilation can result in a change in the blood pH, causing a condition called
    (a) acidosis   (c) alkalosis
    (b) diabetes mellitus   (d) congestive heart failure

25. The urine of a patient suffering from alkalosis will have a pH
    (a) less than 7   (c) greater than 9
    (b) greater than 7   (d) less than 4

26. The best buffer systems are composed of
    (a) a strong acid and its conjugate base
    (b) a weak acid and its conjugate base
    (c) a strong base and its conjugate acid
    (d) a weak base and its conjugate acid

27. Describe the effects of (a) adding acid and (b) adding base to an aqueous solution containing the following buffer system:

$$NH_4OH + H^+ \rightleftharpoons NH_4^+ + H_2O$$

Answers to Self-Test Questions in Chapter 12

1. (a) React with metals to produce hydrogen, taste sour, turn litmus paper red, neutralize bases (b) Taste bitter, feel slippery, turn litmus paper blue, neutralize acids **2.** $acid_1$, $base_2$, $acid_2$, $base_1$ **3.** $base_1$, $acid_2$, $acid_1$, $base_2$ **4.** $acid_1$, $base_2$, $acid_2$, $base_1$ **5.** $base_1$, $acid_2$, $acid_1$, $base_2$ **6.** (a) H_2O, HPO_4^{2-}, H_2CO_3, HNO_3, HCl, H_2SO_4 (b) NO_3^-, H_2O, HPO_4^{2-}, CO_3^{2-}, PO_4^{3-}, OH^-
7. (a) hydrofluoric acid (b) nitric acid (c) potassium dihydrogen phosphate (d) sodium hydrogen sulfite (sodium bisulfite) (e) nitrous acid (f) calcium sulfate

8. (a) $\quad Ca(OH)_2 + 2HCl \longrightarrow 2H_2O + CaCl_2$
$\qquad\quad 2OH^- + 2H^+ \longrightarrow 2H_2O$
$\quad$ (b) $\quad HNO_3 + KOH \longrightarrow H_2O + KNO_3$
$\qquad\quad H^+ + OH^- \longrightarrow H_2O$

9. (a) $[H^+] = 0.1$ M (b) $[OH^-] = \dfrac{1 \times 10^{-14}}{[H^+]} = \dfrac{1 \times 10^{-14}}{1 \times 10^{-1}} = 1 \times 10^{-13}$ M

$\quad$ (c) pH = 1 (d) acidic

10. (a) $[H^+] = 1 \times 10^{-4}$ M (b) $[OH^-] = 1 \times 10^{-10}$ M

$\quad$ (c) pH = 4 $\qquad\qquad$ (d) acidic

11. (a)
$$[H^+] = \frac{1 \times 10^{-14}}{[OH^-]} = \frac{1 \times 10^{-14}}{1 \times 10^{-3}} = 1 \times 10^{-11} \text{ M} \quad \text{(b)} \quad [OH^-] = 0.001 \text{ M}$$

$\quad$ (c) pH = 11 $\qquad$ (d) basic

12. (a) $[H^+] = 1 \times 10^{-8}$ (b) $[OH^-] = 1 \times 10^{-6}$

$\quad$ (c) pH = 8 $\qquad\qquad$ (d) basic

13. (a) $[H^+] = 1 \times 10^{-10}$ (b) $[OH^-] = 1 \times 10^{-4}$

$\quad$ (c) pH = 10 $\qquad\qquad$ (d) basic

14. (a)
$$[H^+] = \frac{36.5 \ \mu g}{100 \text{ mL}} \times \frac{1 \text{ g}}{1 \times 10 \ \mu g} \times \frac{1 \text{ mol}}{36.5 \text{ g}} \times \frac{1000 \text{ mL}}{1 \text{ liter}} = \frac{1 \times 10^{-5} \text{ mol}}{1 \text{ liter}}$$

$\quad$ (b) $[OH^-] = 1 \times 10^{-9}$ (c) pH = 5 $\qquad$ (d) acidic

15. (a) $[H^+] = 1 \times 10^{-12}$ M

(b) $\dfrac{370 \text{ mg Ca(OH)}_2}{1 \text{ liter}} \times \dfrac{1 \text{ g}}{1000 \text{ mg}} \times \dfrac{1 \text{ mol}}{74.1 \text{ g}} = \dfrac{0.005 \text{ mol Ca(OH)}_2}{1 \text{ liter}}$

$[OH^-] = 0.01 \text{ M} = 1 \times 10^{-2} \text{ M}$

(c) pH = 12   (d) basic

16. 50 mL of 0.02 M NaOH contains $\dfrac{0.02 \text{ mol OH}^-}{1000 \text{ mL}} \times 50 \text{ mL}$

$= 0.001 \text{ mol OH}^-$

50 mL of 0.04 M HCl contains $\dfrac{0.04 \text{ mol H}^+}{1000 \text{ mL}} \times 50 \text{ mL}$

$= 0.002 \text{ mol H}^+$

0.001 mole of OH^- will neutralize 0.001 mole of H^+ leaving 0.001 mole of H^+ remaining in 100 mL of solution. Therefore,

(a) $[H^+] = \dfrac{0.001 \text{ mol}}{100 \text{ mL}} \times \dfrac{1000 \text{ mL}}{1 \text{ liter}} = 0.01 \text{ M}$ (b) $[OH^-] = 1 \times 10^{-12} \text{ M}$

(c) pH = 2   (d) acidic

17. $\dfrac{0.20 \text{ mol}}{1000 \text{ mL}} \times 15 \text{ mL} = 0.0030 \text{ mol base} = 0.0030 \text{ mol acid}$

$\dfrac{0.0030 \text{ mol acid}}{100 \text{ mL}} \times \dfrac{1000 \text{ mL}}{1 \text{ liter}} = 0.030 \text{ M acid}$

18. (a) 80.9 g   (b) 40.0 g   (c) 41.0 g   (d) 37.0 g

19. $300 \text{ mL} \times \dfrac{1 \text{ liter}}{1000 \text{ mL}} \times \dfrac{0.150 \text{ Eq HBr}}{1 \text{ liter}} \times \dfrac{80.9 \text{ g}}{1 \text{ Eq}} = 3.64 \text{ g HBr}$

Add 3.64 g of HBr to enough water to make 300 mL of solution.

20. $\dfrac{0.18 \text{ g}}{100 \text{ mL}} \times \dfrac{1000 \text{ mL}}{1 \text{ liter}} \times \dfrac{1 \text{ Eq Ca(OH)}_2}{37.0 \text{ g}} = 0.049 \text{ N Ca(OH)}_2$

21. $50 \text{ mL} \times N_a = 37 \text{ mL} \times 0.15 \text{ N} ; \ N_a = 0.11 \text{ N}$

22. d **23.** a **24.** c **25.** b **26.** b,d **27.** (a) drive the reaction to the right, producing more NH_4^+ and H_2O (b) drive the reaction to the left, producing more NH_4OH and H^+.

Chapter 13   CARBON AND HYDROGEN: SATURATED HYDROCARBONS

This chapter begins our study of the chemistry of carbon compounds. It introduces new vocabulary, new methods for naming compounds, and new ways of writing chemical formulas that describe the structure of these compounds. All this may seem a bit complex at first. But as was true of Chapter 1, a little extra time spent mastering these new concepts will make it much easier to read and study the rest of the book.

13.1   The Role of Carbon

Hydrogen, nitrogen, carbon, and oxygen are the most abundant elements in living organisms, making up 99.3% of all the atoms in your body. This section examines why carbon-containing molecules form the building blocks of life. Carbon is unique among elements on the periodic table in forming strong stable bonds with up to four other carbon atoms. Molecules containing carbon atoms can form long chains, branched chains, and rings. These carbon compounds can form molecules (called isomers) having identical molecular formulas but different geometric structures.

Important Term

    isomer

13.2   Bonding with Hybrid Orbitals   (optional section)

Covalent bonds are formed by the overlap of atomic orbitals: the greater the overlap, the stronger the bond. In forming some bonds, atomic orbitals will mix or hybridize to form hybrid orbitals, whose shapes allow for greater overlap. In order to form four equal covalent bonds, the 2s and 2p orbitals in carbon hybridize and form four equal sp^3 orbitals, each directed toward the corner of a tetrahedron. So, when the carbon atom forms its four single bonds, the resulting molecule is tetrahedral in shape.

Important Terms

    hybridization   hybrid orbital sp^3 orbital

13.3 Organic Chemistry

In addition to forming stable bonds with other carbon atoms,
carbon can form stable bonds with other elements such as
hydrogen, oxygen, nitrogen, sulfur, phosphorus, and the halogens.
This results in an enormous number of different compounds and has
led to a separate field of chemistry, called organic chemistry,
devoted to the study of carbon compounds.

Important Term

    organic chemistry

13.4 Hydrocarbons

Hydrocarbons are a large class of organic compounds whose
molecules contain only hydrogen and carbon. Petroleum is the
major source of hydrocarbon compounds. The different
hydrocarbons in petroleum are separated by a process called
fractional distillation. This process makes use of the fact that
compounds with longer hydrocarbon chains have higher boiling
points than compounds with shorter chains.

Important Terms

    hydrocarbon   petroleum   fractional distillation

13.5 Writing Structural Formulas

Structural formulas are chemical diagrams that show the position
of the atoms in a molecule. Condensed structural formulas show
the position of the atoms without drawing all the bond lines.

Important Terms

    structural formula   condensed structural formula

Example _____

There are two isomers that have the molecular formula C_4H_{10}. We
can write their structural formulas as follows:

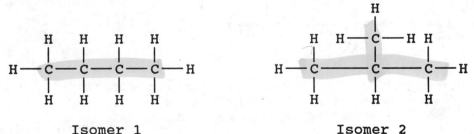

        Isomer 1   Isomer 2

1. Write the condensed structural formula for isomer 1.

$$CH_3CH_2CH_2CH_3 \quad \text{or} \quad CH_3(CH_2)_2CH_3$$

2. Write the condensed structural formula for isomer 2.

$$\begin{array}{c} CH_3 \\ | \\ CH_3CHCH_3 \end{array} \quad \text{or} \quad CH_3CH(CH_3)CH_3$$

Self-Test

For questions 1 to 4, write the condensed structural formula for the given structural formula.

1.

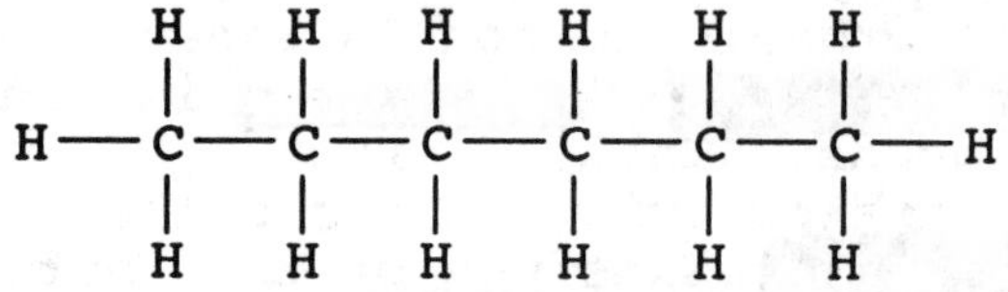

$$H-\underset{\underset{H}{|}}{\overset{\overset{H}{|}}{C}}-\underset{\underset{H}{|}}{\overset{\overset{H}{|}}{C}}-\underset{\underset{H}{|}}{\overset{\overset{H}{|}}{C}}-\underset{\underset{H}{|}}{\overset{\overset{H}{|}}{C}}-\underset{\underset{H}{|}}{\overset{\overset{H}{|}}{C}}-\underset{\underset{H}{|}}{\overset{\overset{H}{|}}{C}}-H$$

$CH_3(CH_2)_4CH_3$

2.

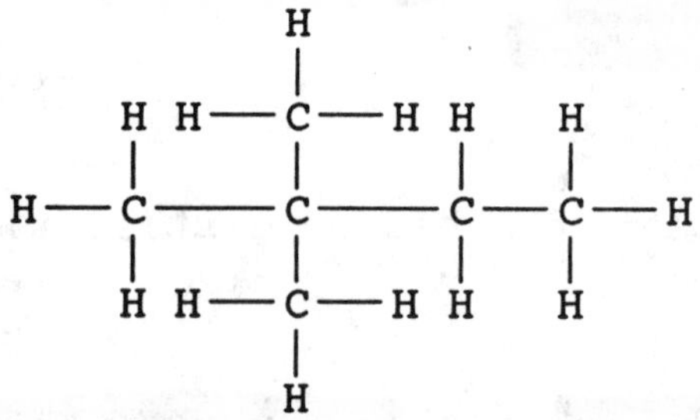

$CH_3C(CH_3)_2CH_2CH_3$

3.

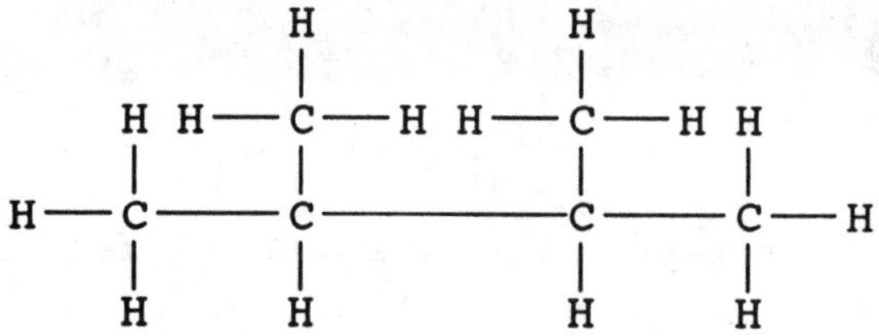

$CH_3CH(CH_3)CH(CH_3)CH_3$

4.

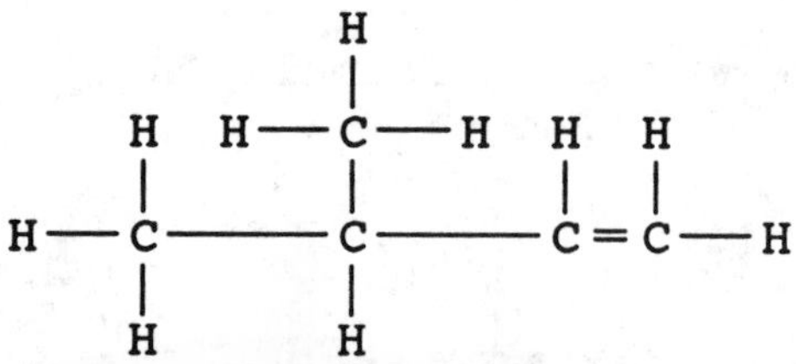

$CH_3CH(CH_3)CH=CH_2$

For questions 5 to 8, write the complete structural formula for the given condensed structural formula.

138

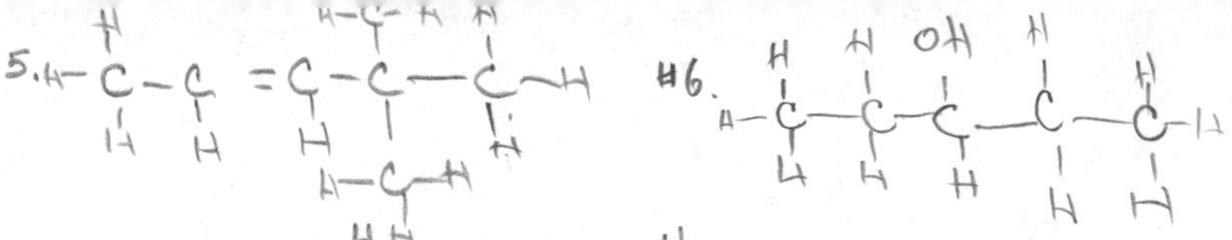

5. $CH_3CH=CHC(CH_3)_2CH_3$

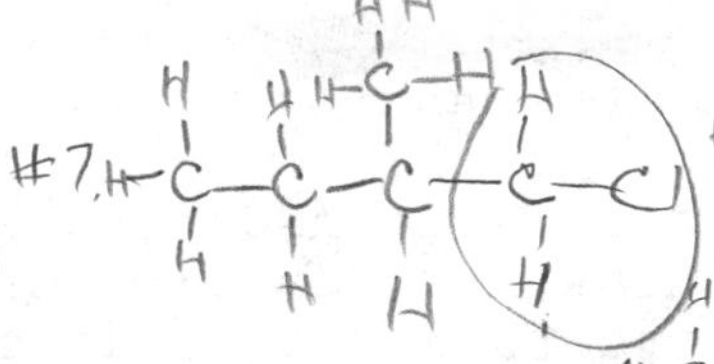

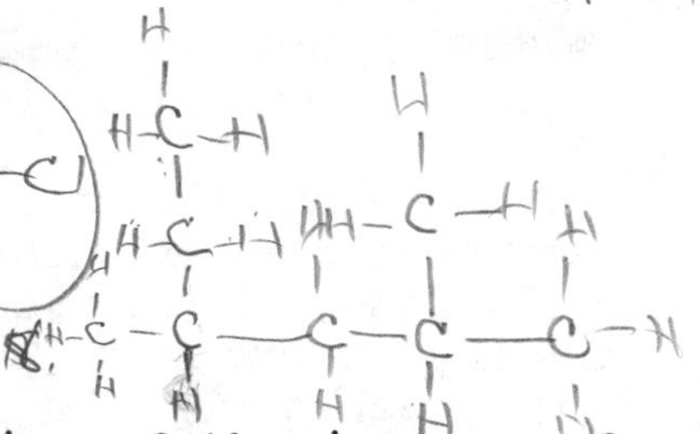

6. $CH_3CH_2CHOHCH_2CH_3$

does it matter where

7. $CH_3CH_2CH(CH_3)CH_2Cl$

the cl is?

8. $CH_3CH(C_2H_5)CH_2CH(CH_3)CH_3$

9. Write the structural formulas for nine of the isomers of C_7H_{16}.

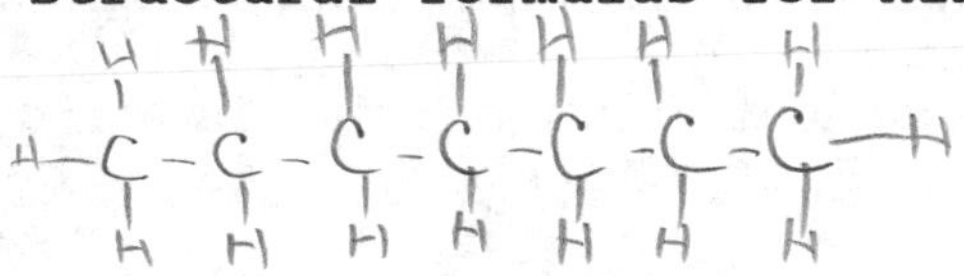

13.6 Alkanes

The alkanes are a class of hydrocarbons whose molecular formulas fit the general formula C_nH_{2n+2}. The chief characteristic of the alkanes is that all the bonds in the compound are single bonds. Compounds that contain all carbon-to-carbon single bonds are said to be saturated. As a result, alkanes are the least reactive class of hydrocarbons.

Important Terms

    alkane   saturated hydrocarbon

13.7 IUPAC Nomenclature

The IUPAC rules give a standardized procedure for naming organic compounds. Study carefully the basic rules listed in Section 13.7 of your text.

Example _____

1. Name the following compound:

$$CH_2CH_3$$
$$|$$
$$CH_3CH_2CHCH_2CH_2CHCH_3$$
$$|$$
$$CH_3$$

heptane
(7) for chain

(a) All the bonds in this compound are single bonds so it is an alkane and the name will end in -ane.

(b) Count the number of carbons in the longest carbon chain. There are seven. The prefix that means seven is hept-. So, this compound is a heptane.

(c) What are the groups that are attached to the carbon

chain? There is a methyl group (–CH$_3$) and an ethyl group (–CH$_2$CH$_3$) attached to the chain.

(d) The name of each attached group must be preceded by a number that indicates its position on the carbon chain. In this case, there are two possibilities: 3-ethyl-6-methyl or 5-ethyl-2-methyl. The name to choose is the one that will give each group the lowest possible number. Therefore, the name of this compound is 5-ethyl-2-methylheptane.

2. What is the IUPAC name for each of the isomers shown in the example in Section 13.5 of the study guide?

Isomer 1. There are four carbon atoms in the main carbon chain, each connected by a single bond. Therefore, this compound is an alkane. The correct prefix is but- (4 carbons), and suffix -ane (alkane). There are no groups other than hydrogen attached to the carbon chain, so the IUPAC name of this compound is butane.

Isomer 2. There are three carbon atoms in the main carbon chain, and this is also an alkane. Therefore, the prefix will be prop-, and the suffix -ane. There is a group attached to the main carbon chain, and it is the methyl group. Since this group is attached to the second carbon, the IUPAC name is 2-methylpropane.

Self-Test _____

10. Name the nine structural isomers of C$_7$H$_{16}$ given in the answer to Self-Test question 9. heptane – 2-methylhexane 3 methylhexane

11. Write the structural formula for each of the following compounds:
    (a) propane
    (b) octane
    (c) 3-ethylnonane
    (d) 3-methylheptane
    (e) 2,2-dichloropentane
    (f) 1,3-dibromopropane
    (g) 4-ethyl-2-methylhexane
    (h) 2,2,3,3-tetramethylbutane

13.8 Constitutional Isomers

Constitutional isomers are compounds having identical molecular formulas, but different three-dimensional arrangements of atoms in their molecules. Although these isomers have the same number of atoms of each element, the atoms are attached differently.

140

This gives the isomers different chemical and physical properties. The number of possible constitutional isomers increases as the number of carbon atoms in the compound increases.

Important Term

constitutional isomer

13.9 Reactions of Alkanes

Although alkanes are fairly unreactive, they do undergo several chemical reactions, including oxidation. Our cells meet their energy needs through oxidation of compounds that are derivatives of alkanes. Oxidation is an exothermic reaction between a compound and oxygen. When the reaction rate is very fast and the heat produced can be felt and seen, we say that combustion is occurring. The products of complete oxidation of an alkane are carbon dioxide and water. If there is an insufficient supply of oxygen present when alkanes burn, incomplete oxidation (incomplete combustion) occurs and the products are carbon monoxide and water. The effects on our bodies of carbon monoxide in polluted air and cigarette smoke are discussed in this section.

Alkanes also undergo substitution reactions with nitric acid and the halogens. In a substitution reaction, an atom or group of atoms is substituted for a hydrogen on the alkane.

Important Terms

oxidation   substitution reaction   combustion

Self-Test _____

For questions 12 to 19, fill in the blank with the correct word or words.

12. A major industrial source of hydrocarbons is *petroleum*, which was formed from *decayed plant materials*

13. The hydrocarbons in this source are separated from one another using a process called *fractional distillation*, which makes use of the fact that short-carbon-chain hydrocarbons have *lower* boiling points than long-chain hydrocarbons.

14. Methane and ethane belong to a class of hydrocarbons called *alkanes*. The general molecular formula for this class of compounds is _____.

C_nH_{2n+2}

141

15. A hydrocarbon that contains all carbon-to-carbon single bonds is *saturated*

16. Compounds that have identical molecular formulas, but different arrangements of atoms in their molecules, are called *constitutional* . *Isomers*

17. Alkanes are relatively *unreactive* (reactive, unreactive), but they do undergo *oxidation* and *substitution* reactions.

18. The products of complete combustion (oxidation) of an alkane are $CO_2 + H_2O$ and the products of incomplete combustion (oxidation) are $CO + H_2O$.

19. The effect of carbon monoxide on our bodies is to *reduce o carry* *position of the blood*. Our bodies counteract this effect by↑ *Resp+Hart* and ↑*Hd Red cells*, both of which increase the risk of *Heart disease*.

20. Complete and balance the equations for the following reactions:

    (a) $CH_3CH_3 + O_2 \longrightarrow$ C_2

    (b) $CH_3CH_3 + HNO_3 \longrightarrow$

Answers to Self-Test Questions in Chapter 13

1. $CH_3CH_2CH_2CH_2CH_2CH_3$ 2. $CH_3C(CH_3)_2CH_2CH_3$ 3. $CH_3CH(CH_3)CH(CH_3)CH_3$
4. $CH_3CH(CH_3)CH=CH_2$

5.

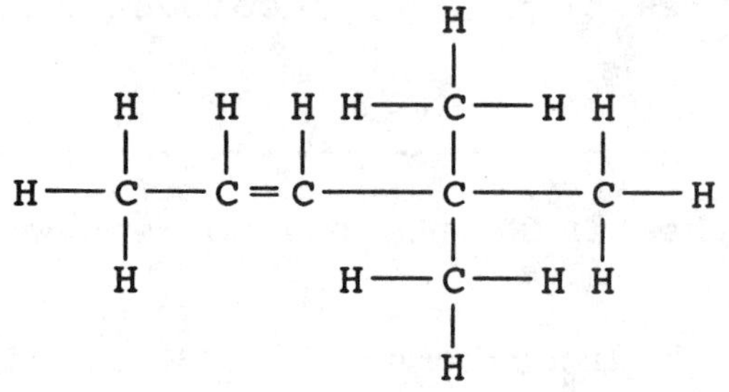

6.

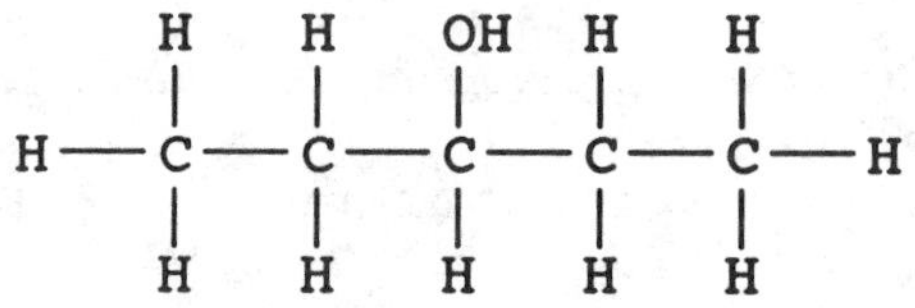

7.

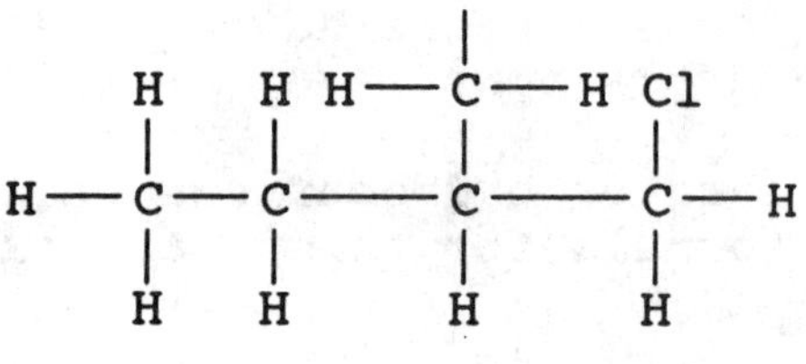

142

8.

```
            H
            |
        H — C — H           H
            |               |
    H   H — C — H   H   H — C — H   H
    |       |       |       |       |
H — C ————— C ————— C ————— C ————— C — H
    |       |       |       |       |
    H       H       H       H       H
```

9. (Note: only the carbon atoms are shown. Each — represents a bond to a hydrogen atom)

1.
```
    |   |   |   |   |   |   |
  — C — C — C — C — C — C — C —
    |   |   |   |   |   |   |
```

2.
```
                |
              — C —
    |   |       |   |   |   |
  — C — C ————— C — C — C — C —
    |   |       |   |   |   |
```

2.
```
            |
          — C —
    |   |   |   |   |   |
  — C — C — C — C — C — C —
    |   |   |   |   |   |
```

4.
```
            |       |
          — C —   — C —
    |   |   |       |   |   |
  — C — C ————— C — C — C —
    |   |   |       |   |   |
```

5.
```
            |           |
          — C —       — C —
    |   |   |   |   |   |
  — C — C — C — C — C —
    |   |   |   |   |   |
```

6.
```
                |
              — C —
    |   |       |   |   |   |
  — C — C ————— C — C — C — C —
    |   |       |   |   |   |
              — C —
                |
```

7.
```
                |
              — C —
    |   |   |   |   |
  — C — C — C — C — C —
    |   |   |   |   |
              — C —
                |
```

8.
```
            |           |
          — C —       — C —
    |   |   |           |   |
  — C — C ————————————— C — C —
    |   |   |           |   |
          — C —
            |
```

9.
```
    |   |   |   |   |
  — C — C — C — C — C —
    |   |   |   |   |
              — C —
                |
              — C —
                |
```

10. (1) heptane   (2) 2-methylhexane   (3) 3-methylhexane
(4) 2,3-dimethylpentane   (5) 2,4-dimethylpentane
(6) 2,2-dimethylpentane   (7) 3,3-dimethylpentane

143

(8) 2,2,3-trimethylbutane (9) 3-ethylpentane

11. (a) $CH_3CH_2CH_3$ (b) $CH_3CH_2CH_2CH_2CH_2CH_2CH_2CH_3$

$$
\begin{array}{c}
CH_2CH_3 \\
|
\end{array}
\qquad\qquad
\begin{array}{c}
CH_3 \\
|
\end{array}
$$

(c) $CH_3CH_2CHCH_2CH_2CH_2CH_2CH_2CH_3$ (d) $CH_3CH_2CHCH_2CH_2CH_2CH_3$

$$
\begin{array}{c}
Cl \\
|
\end{array}
\qquad\qquad
\begin{array}{cc}
Br & Br \\
| & |
\end{array}
$$

(e) $CH_3CCH_2CH_2CH_3$ (f) $CH_2CH_2CH_2$

$$
\begin{array}{c}
| \\
Cl
\end{array}
$$

$$
\begin{array}{cc}
CH_3 & CH_2CH_3 \\
| & |
\end{array}
\qquad\qquad
\begin{array}{cc}
CH_3 & CH_3 \\
| & |
\end{array}
$$

(g) $CH_3CHCH_2CHCH_2CH_3$ (h) $CH_3-C-C-CH_3$

$$
\begin{array}{cc}
| & | \\
CH_3 & CH_3
\end{array}
$$

12. petroleum, decayed plant material **13.** fractional distillation, lower **14.** alkanes, C_nH_{2n+2} **15.** saturated **16.** constitutional isomers **17.** unreactive, oxidation and substitution **18.** CO_2 and H_2O, CO and H_2O **19.** reduce the oxygen carrying capacity of the blood, increase respiration and heart rate, increase the number of red blood cells, heart disease

20. (a) $2CH_3CH_3 + 7O_2 \longrightarrow 4CO_2 + 6 H_2O$

   (b) $CH_3CH_3 + HNO_3 \longrightarrow CH_3CH_2NO_2 + H_2O$

144

Chapter 14   CARBON AND HYDROGEN: UNSATURATED HYDROCARBONS

In this chapter we study hydrocarbons that contain double or triple bonds: the unsaturated hydrocarbons.

14.1   Unsaturated Hydrocarbons

The unsaturated hydrocarbons contain carbon-to-carbon double bonds and carbon-to-carbon triple bonds. Compounds that contain more than one double or triple bond are called polyunsaturated. Double or triple bonds form an unstable spot in the hydrocarbon molecule. As a result, unsaturated hydrocarbons are more reactive than saturated hydrocarbons.

Important Terms

unsaturated   polyunsaturated

14.2   Alkenes

The alkenes are the class of hydrocarbons that contain one or more carbon-to-carbon double bonds. The general formula for alkenes containing one double bond is C_nH_{2n}.

Important Term

alkenes

14.3   Naming Alkenes

To name an alkene, you follow the same rules we discussed for the alkanes except that the name ends in -ene. The main carbon chain must be the one that contains the double bond. As was the case with substituted groups, the position of the double bond is indicated by a number. If there is more than one double bond, a prefix is used to indicate the number of double bonds.

Example _____

1.   Name the following compound

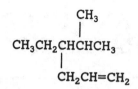

(a) The main carbon chain of an alkene must be the one containing the double bond. It will be easier to determine the name if we rewrite the structure as follows:

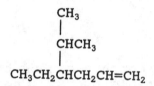

Rewritten as shown, we see that the main carbon chain has six carbons: a hexene. The double bond is attached to the first carbon, so this is 1-hexene.

(b) There is one substituted group: an isopropyl group (see Table 13.2 for the names of such groups). The substituted group is on carbon 4, so the name is 4-isopropyl-1-hexene.

2. Write the structural formula for 3-ethyl-1,3-pentadiene.

(a) The first step in writing a structural formula is to write the carbons in the main carbon chain. The prefix "penta-" tells us that there are five carbons.

C   C   C   C   C

(b) The suffix "-diene" indicates that there are two double bonds, one on carbon 1 and the second on carbon 3. The rest of the bonds between the carbons are single bonds.

C=C-C=C-C

(c) There is one substituted group: an ethyl group (-CH₂CH₃) on carbon 3.

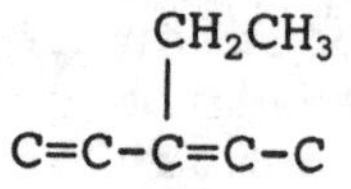

$$CH_2CH_3$$
$$|$$
$$C=C-C=C-C$$

(d) The rest of the bonds on each carbon atom will be attached to hydrogen atoms. When adding hydrogens, we must make sure that each carbon has four bonds attached to it.

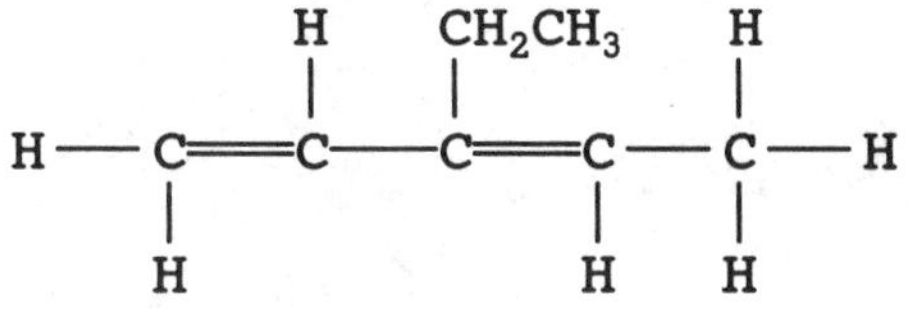

$$H-\underset{\underset{H}{|}}{\overset{\overset{H}{|}}{C}}=\underset{\overset{|}{\underset{|}{\,}}}{\overset{\overset{CH_2CH_3}{|}}{C}}-\underset{\overset{|}{H}}{\overset{\overset{H}{|}}{C}}=\underset{\overset{|}{H}}{\overset{}{C}}-\underset{\overset{|}{H}}{\overset{\overset{H}{|}}{C}}-H \qquad \text{or} \qquad CH_2\!=\!CHC\!=\!CHCH_3$$

with CH_2CH_3 above.

1. Write the structural formula for the following compounds.
   (a) propene   (c) 4-propyl-2,4,6-octatriene
   (b) 2,3-dimethyl-2-butene   (d) 4-ethyl-2-methyl-2-hexene

2. Name the following compounds.

   (a) $CH_3CH\!=\!CHCH_3$ (c) $CH_3CH_2CH_2CH\!=\!CH_2$

   (b) $CH_3-CH\!=\!CH$ (with CH_3 above)   (d) $CH_3CHCH\!=\!CHCH_2CH\!=\!CH_2$ (with CH_2CH_3 above)

14.4 The Pi Bond (optional section)

The double bond actually consists of two bonds: the sigma bond and the pi bond. The sigma bond is formed by the overlap of s, p, or hybrid orbitals along the bond axis. The pi bond is formed by the overlap of p orbitals above and below the bond axis. A triple bond consists of one sigma and two pi bonds.

Important Terms

   sigma (σ) bond   pi (π) bond

14.5 Cis-Trans Isomerism

The double bond in an alkene forms a rigid spot on the molecule, permitting two possible arrangements of atoms around the double bond. This type of isomerism is called stereoisomerism or cis-trans isomerism. The cis-isomer has the specified atoms on the same side of the bond. The trans-isomer has the specified atoms on opposite sides of the bond.

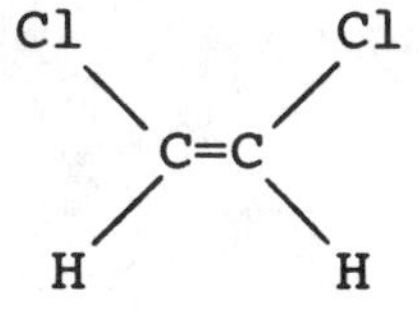

$$\underset{\underset{H}{\diagup}\quad\underset{H}{\diagdown}}{\overset{\overset{Cl}{\diagdown}\quad\overset{Cl}{\diagup}}{C=C}} \qquad\qquad \underset{\underset{H}{\diagup}\quad\underset{Cl}{\diagdown}}{\overset{\overset{Cl}{\diagdown}\quad\overset{H}{\diagup}}{C=C}}$$

      cis-dichloroethene 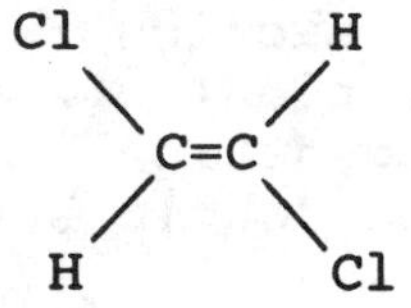   trans-dichloroethene

Important Terms

cis-trans isomerism   stereoisomerism
cis-isomer   trans-isomer

Self-Test

3. Draw and label the cis- and trans-isomers for the following compounds.
   (a) 1,2-difluoropropene   (b) 2-hexene

14.6 Reactions of Alkenes

Alkenes are more reactive than alkanes. They readily undergo oxidation, addition, and polymerization reactions. For our study of organic chemistry, we will modify our definition of oxidation to denote reactions in which one of the reactant molecules gains oxygen atoms or loses hydrogen atoms. Occurring with every oxidation reaction is a reduction reaction. The compound that is reduced will be the one that loses oxygen atoms or gains hydrogen atoms. Alkenes, like alkanes, undergo combustion to form carbon dioxide and water.

An addition reaction occurs when two atoms (or groups of atoms) react with doubly bonded carbon atoms, resulting in the formation of a single bond with one of the reacting atoms (or groups of atoms) bonded to each carbon atom. Water, the halogens, hydrogen, and hydrohalogens (such as HCl) can be added to the double bond. Hydrogenation (the addition of hydrogen to the double bond) is important in the industrial production of margarine.

The process of joining small molecules together to form very large molecules is called polymerization. The small repeating units are called monomers, and the large molecule they form is called a polymer. Polymers occur naturally in living organisms and are produced synthetically. They are the subject of much of Section IV of the textbook.

Important Terms

addition reaction   hydrogenation   hydration
polymerization   monomer   polymer
hydrohalogenation   reduction   oxidation
conjugated double bond

For questions 4 to 10, choose the best answer or answers.

4. Alkenes are
   (a) more reactive than alkanes.   (c) unreactive.
   (b) less reactive than alkanes.   (d) saturated.

5. The addition of water to the double bond of an alkene is called a
   (a) oxidation reaction   (c) hydration reaction
   (b) reduction reaction   (d) hydrogenation reaction

6. The production of margarine involves the addition of this compound to double bonds in vegetable oils.
   (a) water   (c) hydrogen chloride
   (b) hydrogen   (d) oxygen

7. When a reactant molecule gains oxygen atoms, the reaction that has occurred is a
   (a) oxidation reaction   (c) hydrogenation reaction
   (b) reduction reaction   (d) addition reaction

8. When monomer units are joined together to form a very large molecule, the reaction is a
   (a) addition reaction   (c) hydrogenation reaction
   (b) hydration reaction   (d) polymerization reaction

9. Which of the following are addition reactions
   (a) oxidation   (c) hydration
   (b) hydrogenation   (d) polymerization

10. When a reactant molecule gains hydrogen atoms or loses oxygen atoms, the reaction that has occurred is a
    (a) oxidation reaction   (c) hydrogenation reaction
    (b) reduction reaction   (d) addition reaction

11. Draw the structure of the products of the following reactions.

    (a) The polymerization of three molecules of 1-butene

    (b) $CH_3CH=CH_2 + Cl_2 \longrightarrow$

    (c) $CH_3CH=CH_2 \xrightarrow{\text{KMnO}_4}$

    (d) $CH_3CH=CH_2 + H_2 \longrightarrow$

14.7 Alkynes

Alkynes are unsaturated hydrocarbons that contain
carbon-to-carbon triple bonds. The general formula for alkynes
containing one triple bond is C_nH_{2n-2}. Alkynes are named following
the same rules as alkenes except the name ends in -yne.
Acetylene, the simplest alkyne, is used in welding and in the
production of other organic compounds.

Important Term

    alkyne

Self-Test _____

12. What is the molecular formula for an alkyne containing eight
     carbons and one triple bond?

13. Draw the structural formulas for the following compounds:
     (a) propyne   (b) 2-methyl-3-heptyne

14. Name the following compounds:

     (a) $CH_3-C{\equiv}C-CH_3$ (b) $HC{\equiv}C-\underset{\underset{\displaystyle CH_3}{|}}{\overset{\overset{\displaystyle CH_3}{|}}{C}}-CH_2CH_3$

14.8 Cyclic Hydrocarbons

Carbon atoms can form rings containing single, double, and (in
very large rings) triple bonds. The most common cyclic
hydrocarbons contain five or six carbon atoms. This section
discusses the anesthetic properties of cyclic hydrocarbons such
as cyclopropane.

Important Terms

    cyclic hydrocarbon   anesthetic

14.9 Cis and Trans Isomers

The rotation of the carbon atoms around the single bonds in a
cyclic hydrocarbon is restricted. This allows the formation of
cis- and trans-isomers. The cis-isomer of a cyclic compound has
the attached atoms on the same side of the carbon ring. The
trans-isomer has the attached atoms on opposite sides of the
ring.

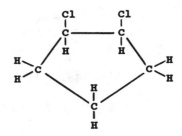

cis-1,2-dichlorocyclopentane

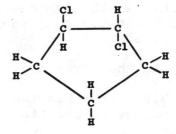

trans-1,2-dichlorocyclopentane

Self-Test

15. Draw the structural formula for the following compounds.
    (a) cyclopentene
    (b) 1,3,5-trichlorocyclohexane
    (c) trans-1,2-dimethylcyclobutane
    (d) 1,3-diethylcyclopentane

14.10 Benzene and Derivatives: Aromatic Hydrocarbons

The aromatic hydrocarbons are a class of cyclic hydrocarbons made up of benzene and its derivatives. Benzene, with a molecular formula of C_6H_6, has an unusual structure often described as a "resonance hybrid". This structure makes the molecule very stable. As a result of its stability, benzene and its derivatives form part of many complex naturally occurring compounds. Be sure to review the schematic ways of writing the structural formulas of cyclic and aromatic hydrocarbons presented in this section.

Important Terms

benzene	resonance hybrid
aromatic hydrocarbon	aliphatic hydrocarbon

14.11 Naming Aromatic Hydrocarbons

Reread carefully the rules for naming aromatic hydrocarbons. You should memorize the names of the compounds given in Rule 1(b).

Example

1. Name the following compound:

This is the benzene derivative called phenol, with a
substituted chlorine in the para or fourth position.
Therefore the name is p-chlorophenol or 4-chlorophenol.

2. Write the structural formula for 3,4-dichlorophenol.

This phenol has two substituted chlorines: one in position 3
and one in position 4.

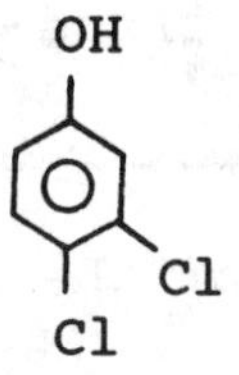

16. Name the following compounds:

(a)

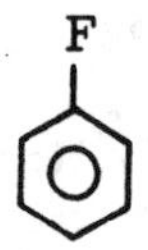

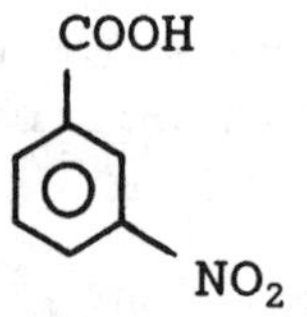

(c)

(b)

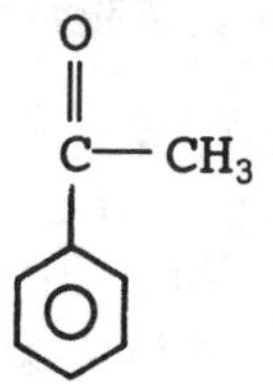

(d)

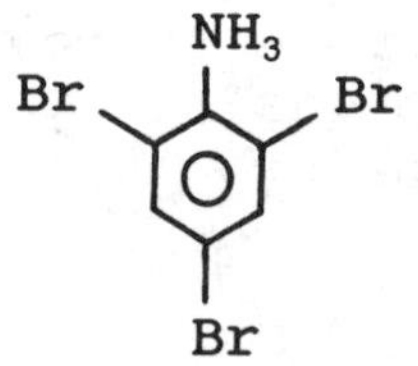

17. Write the structural formulas for the following compounds:
     (a) m-bromophenol
     (b) 2,4-difluorobenzenesulfonic acid
     (c) 1,3-dichlorobenzene
     (d) 2-phenylethanol

14.12 Reactions of Benzene

The structure of benzene makes the ring very stable and resistant
to most chemical changes. The double bonds in the ring, unlike
the double bonds in alkenes, do not undergo addition reactions.
Benzene molecules do undergo reactions that involve the
substitution of an atom or group of atoms for the hydrogens
attached to the ring.

For questions 18 to 21, answer true (T) or false (F).

18. All the carbon-to-carbon bonds in benzene are equal.

19. The double bonds in a benzene ring are less reactive than the double bonds in other alkenes.

20. Animals can synthesize benzene rings.

21. Benzene will readily undergo addition and substitution reactions.

22. Draw the complete structural formulas represented by the following schematic formulas:

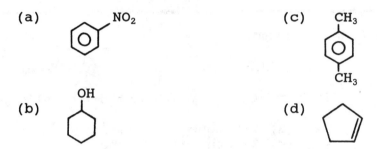

23. Write the equation for the reaction between benzene and chlorine.

14.13   Other Aromatic Hydrocarbons

This section describes a few of the many common compounds containing benzene rings. One compound found in pollutants in the air and in cigarette smoke, 3,4-benzpyrene, is a potent carcinogen. To rid the lungs of this chemical, cells convert it to a polar derivative that is more water soluble and more readily excreted, but also is more carcinogenic.

Halogenated hydrocarbons—compounds in which one or more of the hydrogens on a benzene ring have been replaced by a halogen—are extremely stable and unreactive. This gives these compounds many important uses. But these same properties have led to great concern over the accumulation of these compounds in the environment.

Important Term

   carcinogen

Self-Test _____

For questions 24 to 27, fill in the blank with the correct word or words.

24. Mothballs contain a compound whose structure is composed of two_____.

25. Chemicals that cause cancer are called _____.

26. Our bodies detoxify foreign chemicals such as nonpolar hydrocarbons by _____.

27. PCBs are chemically _____ and dissolve in _____ solvents. This has led to their use as _____ and _____.

Answers to Self-Test Questions in Chapter 14

1. (a) $CH_3CH=CH_2$ (b)

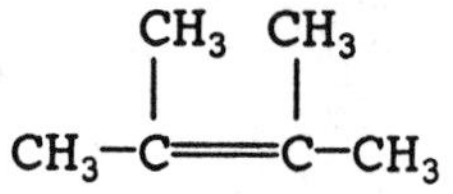

$$CH_3-C \underset{\overset{|}{CH_3}}{\overset{\overset{CH_3}{|}}{=}} C-CH_3$$

(c)

$$CH_3-CH=CH-\underset{\overset{|}{CH_2CH_2CH_3}}{C}=CH-CH=CH-CH_3$$

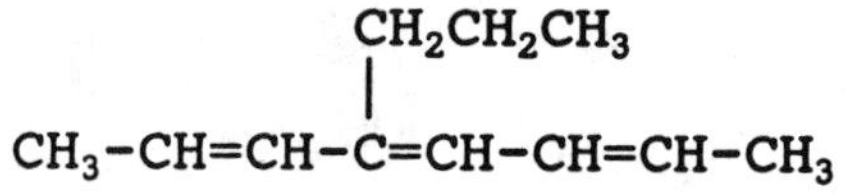

(d)

$$CH_3\underset{\overset{|}{CH_2CH_3}}{\overset{\overset{CH_3}{|}}{C}}=CHCHCH_2CH_3$$

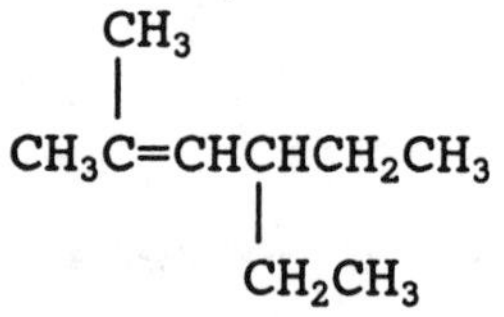

2. (a) 2-butene
(b) 2-methyl-1-propene
(c) 1-pentene
(d) 6-methyl-1,4-octadiene
(check the longest carbon chain)

3. (a)

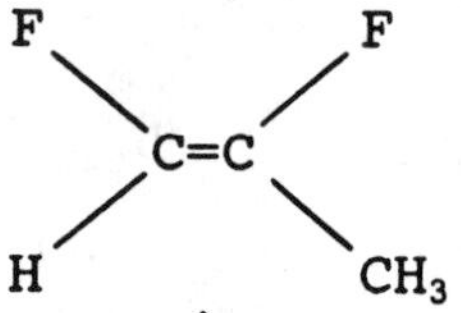

cis 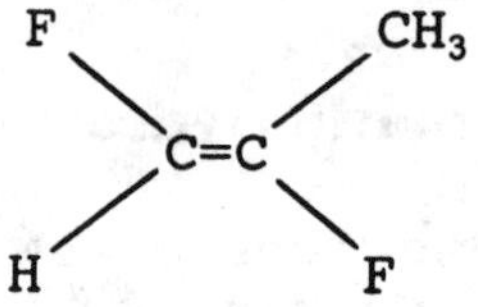 / trans structures

(b)

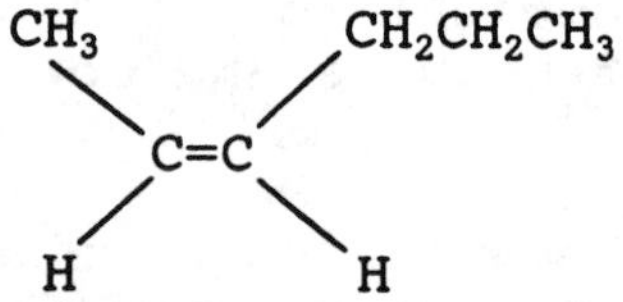

cis 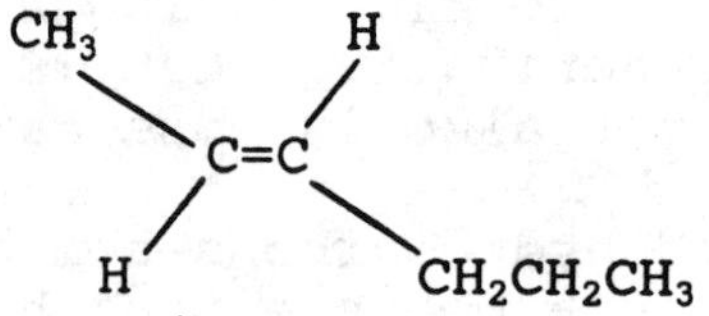 / trans structures

4. a **5.** c **6.** b **7.** a **8.** d **9.** b,c **10.** b

11. (a)

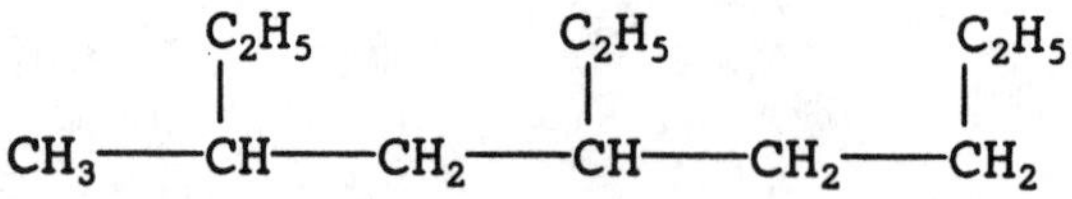

$$CH_3-\underset{\overset{|}{C_2H_5}}{CH}-CH_2-\underset{\overset{|}{C_2H_5}}{CH}-CH_2-\underset{\overset{|}{C_2H_5}}{CH_2}$$

(b)

$$CH_3-\underset{\overset{|}{Cl}}{\overset{\overset{H}{|}}{C}}-\underset{\overset{|}{Cl}}{\overset{\overset{H}{|}}{C}}-H$$

154

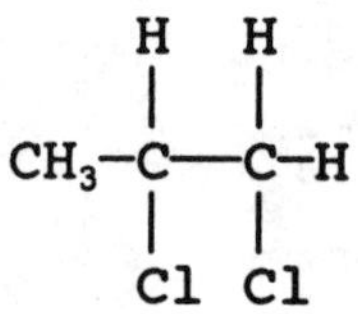

(c) CH₃–CH–CH₂   (d) CH₃CH₂CH₃
          | |
         OH OH

12. C_8H_{14} **13.** (a) CH₃C≡CH   (b)
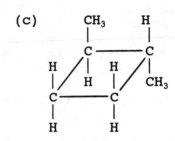
CH₃CHC≡CCH₂CH₂CH₃ (with CH₃ branch)

14. (a) 2-butyne   (b) 3,3-dimethyl-1-pentyne

15. (a)
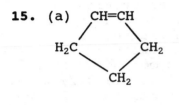

        CH=CH
   H₂C   CH₂
        CH₂

(b)
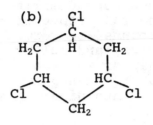

(c)
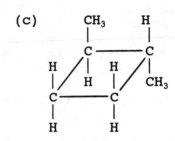

(d)   CH₂CH₃
       |
      CH—CH
  H₂C   HC—CH₂CH₃
      CH₂

16. (a) fluorobenzene
(b) 3-nitrobenzoic acid
(m-nitrobenzoic acid)
(c) acetophenone
(d) 2,4,6-tribromoaniline

17. (a) OH with Br   (b) SO₃H with F, F   (c) Cl, Cl   (d) CH₂CH₂OH

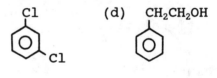

18. T **19.** T **20.** F **21.** F

22. (a)   (b)   (c)   (d)

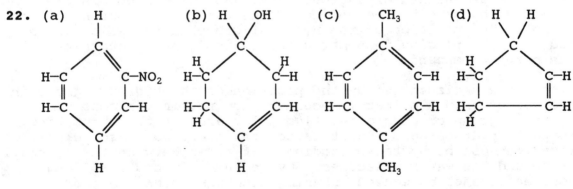

23.

benzene–H + Cl₂ ⟶ benzene–Cl + HCl

24. benzene rings **25.** carcinogens **26.** converting them to more polar, more water-soluble compounds **27.** unreactive, nonpolar, fire retardants, insulators

Chapter 15   ORGANIC COMPOUNDS CONTAINING OXYGEN

The wide variety of hydrocarbons described in the previous chapters only hints at the full range of carbon compounds. Many more types of carbon compounds are formed when elements such as oxygen, nitrogen, phosphorus, and sulfur combine with hydrocarbons. In this chapter we study organic compounds that contain oxygen along with carbon and hydrogen. We classify these compounds by their oxygen-containing functional groups.

15.1 Functional Groups Containing Oxygen

Functional groups are groups of atoms that create a reactive area on an organic molecule and that give the molecule specific chemical properties. In this chapter, we study the functional groups containing oxygen.

Important Term

functional group

15.2 Alcohols

The functional group that makes alcohols a distinct class of compounds is the hydroxyl group, -OH. Specific alcohols are named by changing the -e ending of the name of the parent alkane to -ol. The position of the hydroxyl group in the molecule is indicated by placing a number directly in front of the name of the parent compound.

Alcohols are classified by the position of the hydroxyl group in the molecule into primary, secondary, or tertiary alcohols. The hydroxyl group creates a reactive polar area on the unreactive, nonpolar hydrocarbon. Short-carbon-chain alcohols are soluble in water because of hydrogen bonding that forms between the hydroxyl group and the water molecules. Hydrogen bonds can also form between alcohol molecules, causing alcohols to have higher melting and boiling points than alkanes with comparable molecular weights.

Important Terms

alcohol   hydroxyl group

1. Name the following compound:

$$OH$$
$$|$$
$$CH_3-CH_2-CH-CH-CH_2-CH_3$$
$$|$$
$$CH_2CH_3$$

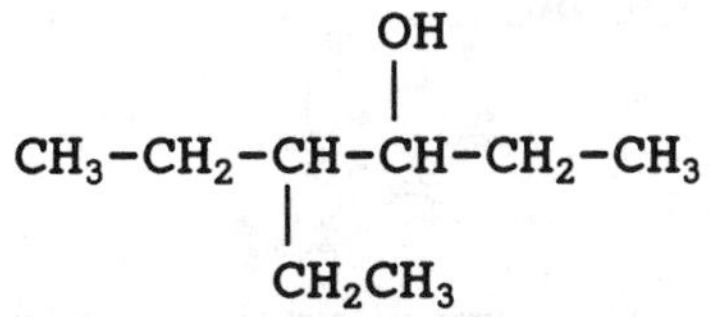

(a) The number of carbon atoms in the longest carbon chain containing the hydroxyl group is 6. Therefore, this is a hexanol.

(b) The hydroxyl group is on carbon 3 (counting from the right so that the hydroxyl group will have the lowest number.) The other substituted group is an ethyl group attached to carbon 4. The complete name for this alcohol is 4-ethyl-3-hexanol.

2. Write the structural formula for 4-chloro-1-pentanol.

(a) The name pentanol indicates that the main carbon chain has five carbons.

$$C-C-C-C-C$$

(b) The hydroxyl group is attached to carbon 1 and a chlorine atom to carbon 4.

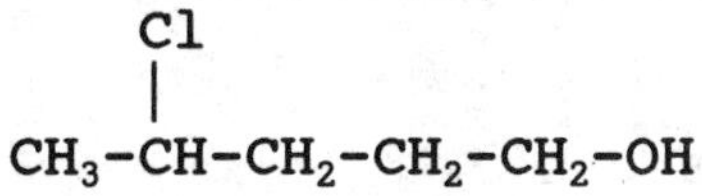

$$Cl$$
$$|$$
$$CH_3-CH-CH_2-CH_2-CH_2-OH$$

1. Write the structural formulas for the following compounds:
    (a) 1-propanol   (c) 2,3-dimethyl-2-butanol
    (b) 2-propanol   (d) 3,4-dibromo-2-hexanol

2. Name the following compounds:

(a)

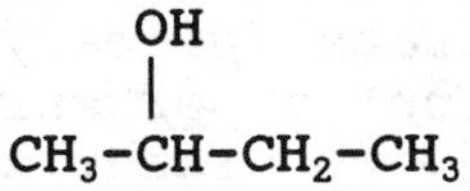

$$OH$$
$$|$$
$$CH_3-CH-CH_2-CH_3$$

(c) $CH_3CH_2CH_2CH_2CH_2OH$

(b)

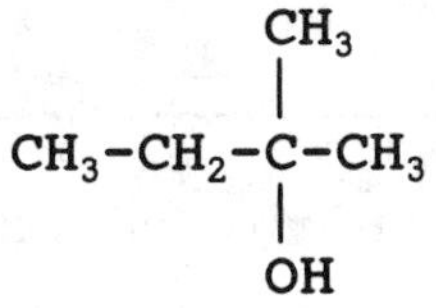

$$CH_3$$
$$|$$
$$CH_3-CH_2-C-CH_3$$
$$|$$
$$OH$$

(d)

$$H$$
$$|$$
$$CH_3-CH-C-H$$
$$| \quad |$$
$$CH_3 \quad OH$$

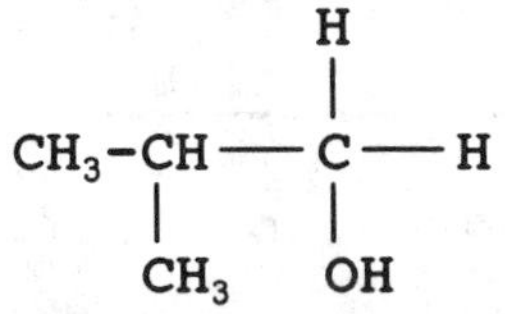

3.	Indicate whether each of the compounds shown in question 2 is a primary, secondary, or tertiary alcohol.

4.	List the following in order of decreasing solubility in water: hexanol, ethanol, butanol.

## 15.3	Straight Chain Alcohols

Of the straight chain alcohols, you are probably most familiar with ethanol and methanol; you should know their structures. Polyhydric alcohols are alcohols that have more than one hydroxyl group.	Important examples are ethylene glycol and glycerol.

Important Term

	polyhydric alcohol

## 15.4	Cyclic and Aromatic Alcohols

Cyclic alcohols have one or more hydroxyl groups substituted for hydrogens on the parent cyclic hydrocarbon.	An aromatic alcohol is formed when hydroxyl groups replace one or more hydrogens on a benzene ring.	The alcohol formed when just one hydrogen is replaced on the benzene ring is phenol.

Important Terms

	cyclic alcohol		aromatic alcohol			phenol

## 15.5	Preparation of Alcohols

Alcohols can be prepared by the addition of water to an alkene (hydration reaction), or by the reduction of an aldehyde or ketone.

## 15.6	Reactions of Alcohols

Alkenes can be produced by the dehydration of an alcohol. Tertiary alcohols are the easiest to dehydrate; primary alcohols are the hardest.	Oxidation (dehydrogenation) of primary alcohols produces aldehydes, and oxidation of secondary alcohols produces ketones.	Tertiary alcohols are not easily oxidized.

Self-Test _____

For questions 5 to 11, fill in the blank with the correct word or words.

5. The formula of wood alcohol is _____.

6. The end products of fermentation are _____.

7. Ethylene glycol is a _____ alcohol. _____ is a trihydric alcohol that is found in body fat.

8. The compound containing one hydroxyl group on a benzene ring is called _____.

9. Alcohols are produced by the _____ of an aldehyde or ketone, or by the _____ of an alkene.

10. The dehydration of propanol produces _____.

11. Aldehydes are produced by the oxidation of _____ alcohols, and ketones by the oxidation of _____ alcohols.

12. Write the formula for the product of the dehydration of each of the following alcohols.

(a) $CH_3CH_2CH_2CH_2OH$

(c) ⬡-CH_2CH_2OH

(b) $CH_3-\overset{\displaystyle OH}{\overset{|}{CH}}-CH_3$

(d) $CH_3-\overset{\displaystyle CH_3}{\overset{|}{CH}}-CH_2-OH$

13. Write the formula for the product of the oxidation of each of the following alcohols.

(a) $CH_3CH_2CH_2OH$

(c) $CH_3-CH_2-\overset{\displaystyle }{CH}-OH$ with CH_3 below

$CH_3-CH_2-\underset{\displaystyle |}{\underset{\displaystyle CH_3}{CH}}-OH$

(b) $CH_3-CH-\underset{\displaystyle |}{\underset{\displaystyle CH_3}{CH}}-CH_3$ with OH on second carbon

$CH_3-\overset{\displaystyle OH}{\overset{|}{CH}}-\underset{\displaystyle |}{\underset{\displaystyle CH_3}{CH}}-CH_3$

(d) $CH_3CH_2\underset{\displaystyle |}{\underset{\displaystyle CH_3}{CH}}CHOH$ with CH_3 below last carbon

$CH_3CH_2\overset{\displaystyle CH_3}{\overset{|}{CH}}\underset{\displaystyle |}{\underset{\displaystyle CH_3}{CH}}OH$

15.7 Ethers

The oxygen in an ether molecule is bonded to two carbon atoms, -C-O-C-. An ether molecule is nonpolar; it is soluble in nonpolar solvents and not soluble in water. Ethers are highly flammable, and many are used as anesthetics. Simple ethers are named by listing both alkyl groups attached to the oxygen.

Important Term

ether

14. Write the structural formulas for the following compounds:
    (a) diethyl ether   (b) methyl propyl ether

15. Name the following compounds:

    (a) $CH_3CH_2CH_2CH_2OCH_2CH_3$ (b) $CH_3-CH-O-CH-CH_3$
                                                            | |
                                                           CH_3 CH_3

15.8 Aldehydes and Ketones

Both aldehydes and ketones contain the carbonyl functional group. Aldehydes have the carbonyl group on the terminal carbon in the molecule, and ketones have the carbonyl group in the middle of the molecule. Aldehydes and ketones are highly reactive compounds, with aldehydes being the more reactive.

Aldehydes are named by adding an -al ending to the name of the longest carbon chain containing the carbonyl group. The names of ketones have an -one ending, and use a number before the name to indicating the position of the carbonyl group.

Important Terms

$$O$$
$$\parallel$$
carbonyl group (-C-)   aldehyde   ketone

15.9 Important Aldehydes and Ketones

Many aldehydes and ketones have pleasant odors and tastes, and are used in making perfumes and flavorings. Formaldehyde is a well-known aldehyde, and acetone is a widely used ketone.

15.10 Preparation of Aldehydes and Ketones

Aldehydes are produced by the oxidation of primary alcohols. Ketones are produced by the oxidation of secondary alcohols.

15.11 Reactions of Aldehydes and Ketones

Reduction (or hydrogenation) of aldehydes and ketones is carried out in the presence of a catalyst such as platinum. The reduction of an aldehyde produces a primary alcohol, and the reduction of a ketone a secondary alcohol. Aldehydes are easily oxidized to form carboxylic acids, but ketones can be oxidized only under extreme chemical conditions.

When aldehydes are dissolved in alcohol, some of the molecules interact to form hemiacetals. A similar molecule, called a hemiketal, is formed when a ketone is dissolved in an alcohol. Internal hemiacetals or hemiketals can form when an aldehyde or ketone group interacts with an alcohol group on the same molecule.

Important Terms

      hydrogenation   hemiacetal   hemiketal

Example

Write the structure of the hemiacetal formed when propanal is dissolved in ethanol.

$$\underset{\text{propanol}}{CH_3CH_2\overset{\overset{\textstyle O}{\|}}{C}H} + \underset{\text{ethanol}}{HOCH_2CH_3} \longrightarrow CH_3CH_2-\underset{\underset{\textstyle OCH_2CH_3}{|}}{\overset{\overset{\textstyle OH}{|}}{C}}-H$$

Self-Test

16. Name the following compounds, and indicate whether the compound is an aldehyde or a ketone.

(a) $CH_3CH_2CH_2-\overset{\overset{\textstyle O}{\|}}{C}-H$

(b) $CH_3CH_2-\overset{\overset{\textstyle O}{\|}}{C}-CH_3$

(c) $CH_3CH_2CH_2-\overset{\overset{\textstyle O}{\|}}{C}-CH_2CH_3$

(d) $Cl-CH_2CH_2CH_2CH_2-\overset{\overset{\textstyle O}{\|}}{C}-H$

17. Write the structural formula of the main product of the following reactions:

(a) $CH_3-\overset{\overset{\textstyle O}{\|}}{C}-H \ + \ H_2 \ \longrightarrow$

(b) $CH_3CH_2CH_2-\overset{\overset{\textstyle O}{\|}}{C}-H \ \overset{KMnO_4}{\longrightarrow}$

(c) $CH_3CH_2-\underset{\underset{\displaystyle H}{|}}{\overset{\overset{\displaystyle OH}{|}}{C}}-CH_3$ $\xrightarrow{KMnO_4}$

(d) $CH_3\underset{\underset{\displaystyle}{|}}{\overset{\overset{\displaystyle CH_3}{|}}{C}}HCH_2CH_2OH$ $\xrightarrow{K_2CrO_7}$

(e) $CH_3CH_2-\overset{\overset{\displaystyle O}{\|}}{C}-CH_2CH_3$ + H_2 $\xrightarrow{Pt}$

(f) $CH_3CH_2CH_2-\overset{\overset{\displaystyle O}{\|}}{C}-H$ + CH_3OH $\longrightarrow$

(g) $CH_3-\overset{\overset{\displaystyle O}{\|}}{C}-CH_3$ + CH_3CH_2OH $\longrightarrow$

15.12   Carboxylic Acids

Carboxylic acids are organic acids containing the carboxyl or carboxylic acid functional group. Carboxylic acids are weak acids. They are named by adding the suffix -oic acid to the name of the longest carbon chain containing the carboxyl group. Because the carboxyl group is polar, hydrogen bonds can form between carboxylic acid molecules, and between carboxylic acid molecules and water. Carboxylic acids, therefore, have higher boiling points than comparable alcohols and are more soluble in water.

Important Terms

carboxylic acid   carboxyl group $(-\overset{\overset{\displaystyle O}{\|}}{C}-OH)$

Self-Test

18.   Name the following compounds:

(a) $CH_3CH_2CH_2CH_2\overset{\overset{\displaystyle O}{\|}}{C}-OH$

162

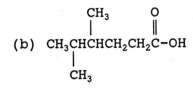

(b) $CH_3CHCHCH_2CH_2C-OH$
with CH_3 above and CH_3 below, and O double bond above the C.

19. Write the structural formula for each of the following compounds:
    (a) propanoic acid   (b) 3-iodobutanoic acid

15.13 Important Carboxylic Acids

Formic acid, which causes the irritation of insect bites, is produced in the liver during the oxidation of methanol. In sufficient quantities, formic acid can cause a severe metabolic disturbance called acidosis. Such acidosis is the cause of death from methanol poisoning. Acetic acid gives the characteristic taste to vinegar, and is used by cells to synthesize fatty acids. Other important carboxylic acids mentioned in this section are benzoic acid, salicylic acid, and oxalic acid.

15.14 Preparation of Carboxylic Acids

Carboxylic acids are prepared by oxidizing primary alcohols or aldehydes with strong oxidizing agents.

Self-Test

For questions 20 to 29, fill in the blank with the correct word or words.

20. The functional group found in carboxylic acids is
    (a) carbonyl group   (c) hydroxyl group
    (b) carboxyl group   (d) amide group

21. When placed in water, propionic acid forms
    (a) the propionate ion   (c) the hydroxide ion
    (b) hydrogen propionate   (d) the hydronium ion

21. The strength of organic acids are comparable to
    (a) nitric acid   (c) sulfuric acid
    (b) carbonic acid   (d) hydrochloric acid

23. The stings of bees contain the carboxylic acid called
    (a) butyric acid   (c) formic acid
    (b) acetic acid   (d) lactic acid

24. Methanol is first oxidized in the liver to this compound which can cause blindness.
    (a) formic acid   (c) acetone
    (b) formaldehyde   (d) acetic acid

25. Vinegar contains
    (a) acetic acid   (c) butyric acid
    (b) formic acid   (d) lactic acid

26. The sodium salt of this acid is a commonly used food preservative.
    (a) lactic acid   (c) capric acid
    (b) tartaric acid   (d) benzoic acid

27. Kidney stones often form from the calcium salt of
    (a) lactic acid   (c) oxalic acid
    (b) oleic acid   (d) benzoic acid

28. Oxidation of 1-propanol by $KMnO_4$ will produce
    (a) propanal   (c) propanone
    (b) propanoic acid   (d) propanoate

29. Oxidation of butanal by $KMnO_4$ will produce
    (a) butanoic acid   (c) 1-butanol
    (b) butanone   (d) pentanoic acid

15.15 Esters

Esters contain the ester functional group. They are derivatives of carboxylic acids. Esters are produced by the esterification (condensation) reaction between an alcohol and an acid. Esters have distinctive flavors and tastes. Esters are named by first listing the alcohol name with a -yl ending, and then the acid name with an -ate ending.

$$CH_3\overset{O}{\overset{\|}{C}}\text{-OH} \quad + \quad CH_3CH_2CH_2OH \quad \longrightarrow \quad CH_3\overset{O}{\overset{\|}{C}}OCH_2CH_2CH_3$$

acetic acid   propanol   propyl acetate

Important Terms

ester   ester group ($-\overset{O}{\overset{\|}{C}}-O-$)
esterification   condensation reaction

15.16 Important Esters

In addition to being used in dynamite, nitroglycerin (an ester of glycerol and nitric acid) has therapeutic value in the treatment of heart disorders. The esters of salicylic acid have many uses. Two examples discussed in this section are methyl salicylate and acetylsalicylic acid (aspirin). Much of the clothing that we wear contains fibers made of polyesters, which are polymers of esters.

Important Term

polyester

15.17 Reactions of Esters

Esters are prepared by a condensation reaction between an alcohol and an acid. Esters can be broken apart by the reverse of this reaction, called hydrolysis. In hydrolysis, a molecule of water is added to the ester linkage, breaking it apart to form an alcohol and an acid. Saponification is the breaking of an ester linkage by a strong base. The products of the saponification of an ester are an alcohol and the salt of the acid.

Important Terms

hydrolysis   saponification

Self-Test

30. Name the following esters:

(a) $CH_3CH_2CH_2CH_2CH_2\overset{\displaystyle O}{\overset{\displaystyle \|}{C}}OCH_2CH_3$

(b) $CH_3CH_2CH_2O\overset{\displaystyle O}{\overset{\displaystyle \|}{C}}H$

31. Write the structural formulas for the following compounds:
     (a) isopropyl benzoate   (b) octyl heptanoate

For questions 32 to 38, fill in the blank with the correct word or words.

32. Esters are formed by a _____ reaction between an _____ and an _____ .

33. The reaction in question 32 produces an ester and _____ .

34. Many esters give distinctive flavors to _____ .

35. Methyl salicylate is formed by the condensation reaction between _____ and _____ . The common name of methyl salicylate is _____ .

36. Aspirin is formed by the condensation reaction between _____ and _____ .

37. An analgesic is a _____ , and an antipyretic is a _____ .

38. Many synthetic clothing fibers are made of _____ .

For questions 39 to 43, write the equation for the reaction.

39. The hydrolysis of methyl salicylate.

40. The condensation reaction between acetic acid and ethanol.

41. The condensation reaction between benzoic acid and 1-butanol.

42. The hydrolysis of ethyl butanoate.

43. The saponification of t-butyl propanoate by potassium hydroxide.

Answers to the Self-Test Questions for Chapter 15

1. (a) $CH_3CH_2CH_2OH$ (b) $CH_3CHOHCH_3$

(c)
$$CH_3 \!-\! \underset{\underset{CH_3}{|}}{\overset{\overset{CH_3}{|}}{CH}} \!-\! \underset{\underset{CH_3}{|}}{\overset{\overset{OH}{|}}{C}} \!-\! CH_3$$

(d)
$$CH_3 \!-\! CH_2 \!-\! CH \!-\! \underset{\underset{Br}{|}}{CH} \!-\! \underset{\underset{Br}{|}}{\overset{\overset{OH}{|}}{CH}} \!-\! CH_3$$

2. (a) 2-butanol (b) 2-methyl-2-butanol (c) 1-pentanol
(d) 2-methyl-1-propanol 3. (a) secondary (b) primary
(c) tertiary (d) primary 4. ethanol, butanol, hexanol 5. CH_3OH
6. ethanol, carbon dioxide, energy 7. dihydric, glycerol
8. phenol 9. reduction, hydration 10. propene 11. primary,
secondary

12. (a) $CH_3CH_2CH{=}CH_2$ (b) $CH_2{=}CHCH_3$ (c) $\langle\!\bigcirc\!\rangle{-}CH{=}CH_2$ (d) $CH_3\underset{\underset{CH_3}{|}}{C}{=}CH_2$

13. (a) $CH_3CH_2\overset{\overset{O}{\|}}{C}H$ (b) $CH_3\overset{\overset{O}{\|}}{C}\underset{\underset{CH_3}{|}}{C}HCH_3$ (c) $CH_3CH_2\overset{\overset{O}{\|}}{C}CH_3$ (d) $CH_3CH_2\underset{\underset{CH_3}{|}}{C}H\overset{\overset{O}{\|}}{C}CH_3$

14. (a) $CH_3CH_2OCH_2CH_3$ (b) $CH_3OCH_2CH_2CH_3$
15. (a) butyl ethyl ether (b) diisopropyl ether
16. (a) butanal, aldehyde (b) butanone, ketone (c) 3-hexanone,
ketone (d) 5-chloropentanal, aldehyde

17. (a) CH_3CH_2OH (b) $CH_3CH_2CH_2\overset{\overset{O}{\|}}{C}OH$ (c) $CH_3CH_2\overset{\overset{O}{\|}}{C}CH_3$

(d) $CH_3\underset{\underset{CH_3}{|}}{C}H CH_2\overset{\overset{O}{\|}}{C}H$ (e) $CH_3CH_2\underset{\underset{OH}{|}}{C}HCH_2CH_3$ (f) $CH_3CH_2CH_2\underset{\underset{OCH_3}{|}}{\overset{\overset{OH}{|}}{C}}{-}H$

(g) $CH_3 \!-\! \underset{\underset{OCH_2CH_3}{|}}{\overset{\overset{OH}{|}}{C}} \!-\! CH_3$

18. (a) pentanoic acid
(b) 4,5-dimethyl hexanoic acid

19. (a) CH_3CH_2COH (with $\overset{O}{\|}$) (b) CH_3CHCH_2COH (with I and $\overset{O}{\|}$) **20.** b **21.** a,d **22.** b **23.** c
24. b **25.** a **26.** d **27.** c **28.** b **29.** a **30.** (a) ethyl hexanoate
(b) propyl formate

31. (a) 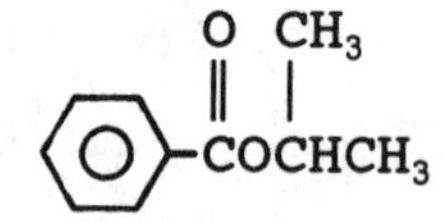 phenyl-$\overset{O}{\overset{\|}{C}}O\overset{CH_3}{\overset{|}{C}}HCH_3$ (b) $CH_3(CH_2)_7O\overset{O}{\overset{\|}{C}}(CH_2)_5CH_3$

32. esterification (condensation), carboxylic acid, alcohol
33. water **34.** fruits **35.** methanol, salicylic acid, oil of
wintergreen **36.** salicylic acid, acetic acid **37.** pain reliever,
fever reducer **38.** polyester

39.

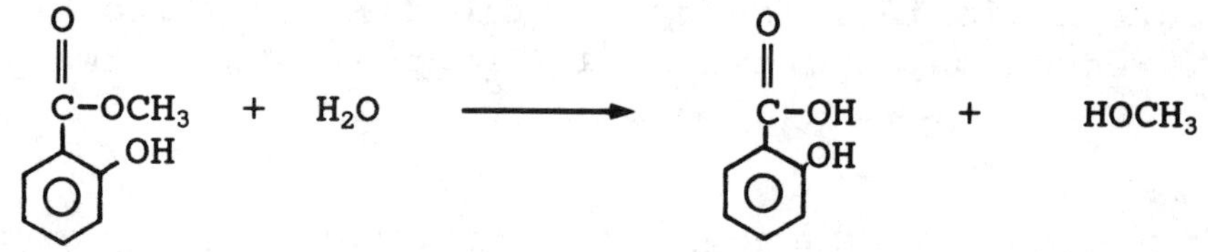

40. $CH_3\overset{O}{\overset{\|}{C}}OH$ + $HOCH_2CH_3$ $\longrightarrow$ $CH_3\overset{O}{\overset{\|}{C}}OCH_2CH_3$ + H_2O

41.

42. $CH_3CH_2CH_2\overset{O}{\overset{\|}{C}}OCH_2CH_3$ + H_2O $\longrightarrow$ $CH_3CH_2CH_2\overset{O}{\overset{\|}{C}}OH$ + $HOCH_2CH_3$

43. $CH_3-\overset{CH_3}{\underset{CH_3}{\overset{|}{\underset{|}{C}}}}-O-\overset{O}{\overset{\|}{C}}-CH_2-CH_3$ + KOH $\longrightarrow$ $CH_3-\overset{CH_3}{\underset{CH_3}{\overset{|}{\underset{|}{C}}}}-OH$ + $CH_3CH_2\overset{O}{\overset{\|}{C}}OK$

Chapter 16   ORGANIC COMPOUNDS CONTAINING NITROGEN

In Chapter 15 we discussed the oxygen-containing functional groups. In this chapter we turn our attention to functional groups that contain nitrogen. Some nitrogen-containing functional groups give basic qualities to the molecule. Nitrogen can also join carbon atoms in ring structures. Nitrogen-containing rings are found in the complex molecules belonging to the class of compounds called alkaloids, which have many physiological effects and are used widely in drug therapy. We discuss a few of these alkaloids at the end of this chapter.

16.1   Functional Groups Containing Nitrogen

Nitrogen is a member of group VA on the periodic table, and is the fourth most abundant element in the human body. Nitrogen can form three single covalent bonds, a double and a single covalent bond, or a triple covalent bond to become stable. It is found in many different arrangements of atoms in organic molecules.

16.2   Amines

Amines are compounds containing the amino functional group. They have very strong odors, and are among the products produced by the decay of dead organisms. Amines can be classified by the number of carbons bonded to the nitrogen, forming primary, secondary, or tertiary amines. Amines are polar compounds; primary and secondary amines can form hydrogen bonds with other amines or water. As a result, amines with short carbon chains are soluble in water.

Amines are named by listing each of the groups attached to the nitrogen and adding the suffix -amine. In more complicated molecules, the -NH₂ group is identified by the term -amino. Any groups attached to the amino group are identified by an "N" before the name of the attached group.

Important Terms

    amine   amino group

1. Name the following compound:

    $$CH_3-CH_2-NH-CH_2-CH_3$$

    There are two ethyl groups attached to the nitrogen.
    Therefore, the name is diethylamine.

2. Write the structural formula for 3-(N-methylamino)-butanal.

    Butanal is an aldehyde with four carbon atoms.

$$
\begin{array}{c}
\quad\quad\quad O \\
\quad\quad\quad \| \\
C-C-C-C-H
\end{array}
$$

The "3-(N-methylamino)" indicates that there is an amino
group on carbon 3, and the amino group has a methyl group
substituted for one of the hydrogens. The complete
structural formula is:

$$
\begin{array}{c}
CH_3-NH \quad\quad O \\
\quad\quad | \quad\quad\quad \| \\
CH_3-CH-CH_2-C-H
\end{array}
$$

16.3 Basic Properties of Amines

The unshared pair of electrons on the nitrogen in an amine allows
the nitrogen to share electrons with a hydrogen ion. By
accepting hydrogen ions, the amine can act as a base. Amines are
weak bases. The specific group attached to the nitrogen atom
will affect the strength of the basic properties of the amine.

16.4 Reactions of Amines

Amines will react with acids to form organic salts. When heated
with alkyl halides, they form quaternary ammonium salts.
Secondary amines will react with nitrites to form nitrosamines,
compounds that are carcinogenic. Nitrosamines may form in the
stomach from nitrites and secondary amines in food.

Important Terms

> quaternary ammonium salt   coordinate covalent bond
> donor-acceptor bond

1. Draw the structural formula for each of the following compounds:
   - (a) hexylamine
   - (b) dibutylamine
   - (c) methylethylpropylamine
   - (d) 2-(N-methylamino)-1-propanol

2. Indicate whether each of the compounds in question **1** is a primary, secondary, or tertiary amine.

3. Name each of the following compounds:

   (a) NH_3

   (b) 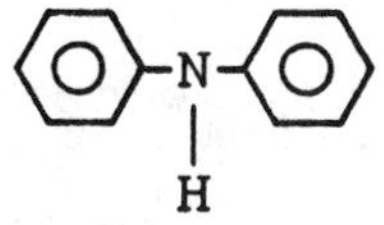 ![benzene-N(H)-benzene structure]

   (c) $CH_3CH_2CH_2CH_2NH_2$

   (d) $CH_3CH_2CH_2-N-CH_2CH_2CH_3$ with CH_2CH_3 group on N

4. List the compounds in question 3 in order of increasing basic strength.

5. Predict the products of the following reactions:

   (a) $CH_3CH_2CH_2NH_2 \ + \ HCl \longrightarrow$

   (b) $CH_3CH_2NH \ + \ H_2O \longrightarrow$ with CH_3 group on N

   (c) $(CH_3CH_2)_3N \ + \ CH_3CH_2Br \longrightarrow$

16.5 Amides

Amides are derivatives of carboxylic acids that contain the amide functional group. The bond between the carbon and the nitrogen in the amide group is called an amide linkage and is a very stable bond. The polar amide group is able to form hydrogen bonds. As a result, short-carbon-chain amides are soluble in water, and have higher melting and boiling points than alkanes with comparable molecular weights.

Simple amides with unsubstituted nitrogens (nitrogens attached to two hydrogens) are named by changing the ending of the name of the parent acid to -amide. If there are groups attached to the nitrogen, they are identified by placing an "N" before the name.

Important Terms

      amide   amide linkage

1. Name the following amide:

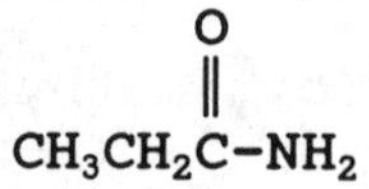

$$CH_3CH_2C\overset{\overset{\displaystyle O}{\|}}{}-NH_2$$

This is a simple amide whose parent acid had three carbons. Therefore, the name is propanamide.

2. Draw the structure of N-ethylbutanamide.

The parent acid is butanoic acid which has 4 carbons.

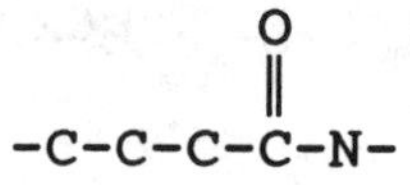

$$-C-C-C-\overset{\overset{\displaystyle O}{\|}}{C}-N-$$

The name tells us that there is one group attached to the nitrogen and it is an ethyl group.

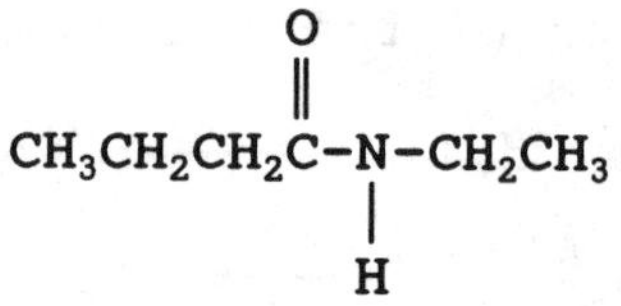

$$CH_3CH_2CH_2\overset{\overset{\displaystyle O}{\|}}{C}-\underset{\underset{\displaystyle H}{|}}{N}-CH_2CH_3$$

16.6 Important Amides

Urea is synthesized from ammonia produced in the breakdown of nitrogen-containing compounds in our bodies. Acetaminophen (an effective analgesic and antipyretic) and the hallucinogen LSD are both amides.

16.7 Preparation of Amides

Amides are prepared by reacting an ester with ammonia or a primary or secondary amine.

Important Term

   ammonolysis

16.8 Reactions of Amides

The amide linkage is very stable, but under certain conditions it can be hydrolyzed or split apart by water. Hydrolysis of an unsubstituted amide produces a carboxylic acid and ammonia. Amides with groups other than hydrogen attached to the nitrogen hydrolyze to form an acid and an amine. In the presence of strong dehydrating agents, amides react to form nitriles.

Important Terms

hydrolysis   dehydration

Self-Test

6.   Which of the following compounds are amides?

(a) 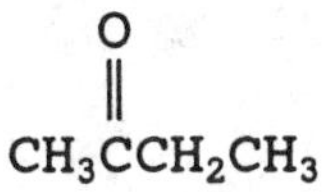 $CH_3\overset{O}{\overset{\|}{C}}CH_2CH_3$

(d) $CH_3CH_2\overset{O}{\overset{\|}{C}}NHCH_2CH_3$

(b) $CH_3CH_2NHCH_2OH$

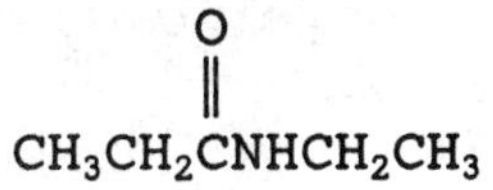

(e) 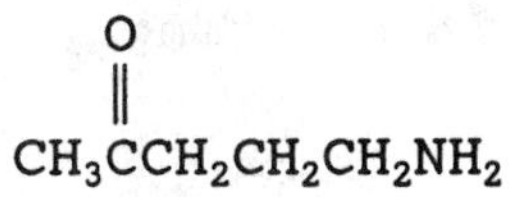 $CH_3\overset{O}{\overset{\|}{C}}CH_2CH_2CH_2NH_2$

(c) $CH_3CH_2\overset{O}{\underset{\underset{CH_3}{|}}{\overset{\|}{C}}}HCNH_2$

(f) 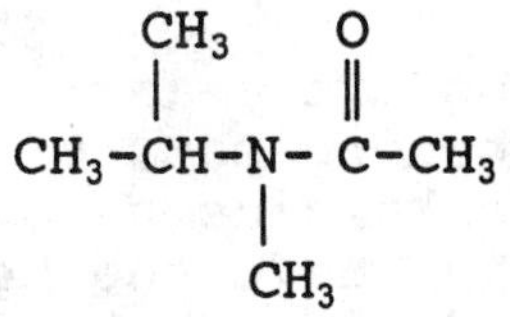 $CH_3-\overset{\overset{CH_3}{|}}{C}H-N-\overset{O}{\overset{\|}{C}}-CH_3$ with CH_3 on N

7.   Name each of the amides in question 6.

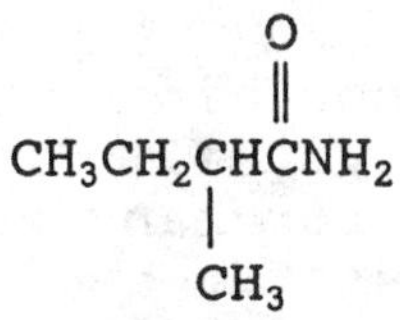

8.   Write the structural formulas for the following compounds:
     (a)   benzamide
     (b)   N-ethyloctanamide
     (c)   N-methyl-N-ethylhexanamide

9.   What are the products of the following reactions?

(a) 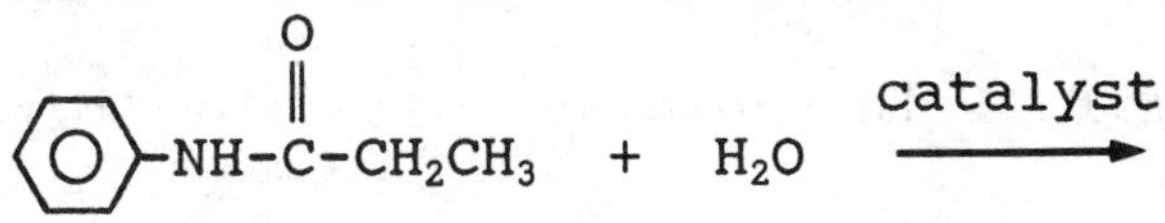 $\langle O \rangle$-NH-$\overset{O}{\overset{\|}{C}}$-CH$_2CH_3$ + H$_2$O $\xrightarrow{\text{catalyst}}$

(b) $CH_3CH_2\overset{O}{\overset{\|}{C}}-NH_2 \xrightarrow{P_4O_{10}}$

(c) $CH_3CH_2\overset{O}{\overset{\|}{C}}-NH_2$ + H$_2$O $\xrightarrow{\text{catalyst}}$

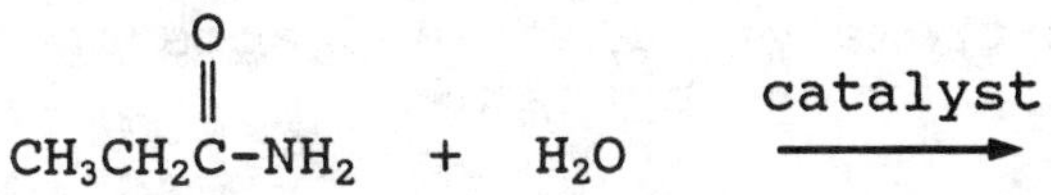

(d) 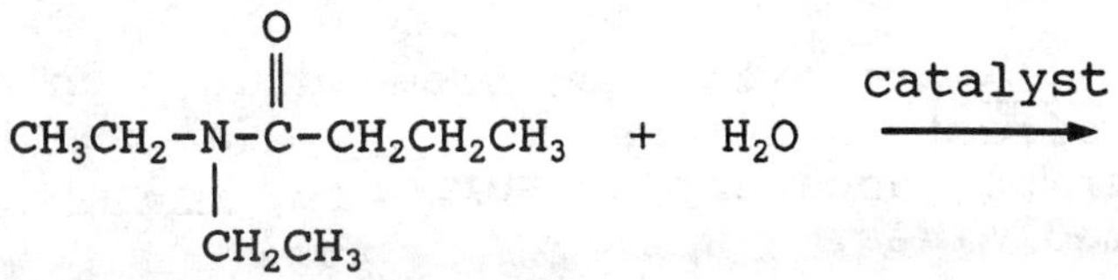 $CH_3CH_2-\underset{\underset{CH_2CH_3}{|}}{N}-\overset{O}{\overset{\|}{C}}-CH_2CH_2CH_3$ + H$_2$O $\xrightarrow{\text{catalyst}}$

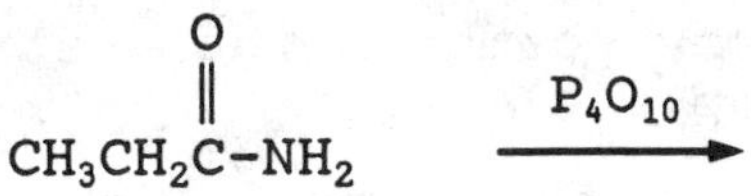

10. Explain with words or a diagram how hydrogen bonding can form between unsubstituted amides such as acetamide, but not between N,N-disubstituted amides such as N,N-dimethylacetamide.

16.9 Heterocyclic Compounds

Rings that contain other elements in addition to carbon and hydrogen are called heterocyclic rings. The number of naturally occurring heterocyclic compounds is very large. We discuss some of these heterocyclic compounds in the next section of the text.

Important Term

   heterocyclic ring

16.10 Alkaloids

The alkaloids are a large class of nitrogen-containing compounds generally forming part of a plant's defense system. Plants containing alkaloids have been used as drugs for many centuries. Many alkaloids have the beneficial properties of reducing pain or producing sedation, but some cause addiction and tolerance on extended use. This section discusses important examples of alkaloids such as epinephrine, norepinephrine, nicotine, caffeine, the opiates, and the barbiturates.

Important Terms

   alkaloid   tolerance   addiction

Self-Test _____

For questions 11 to 18, fill in the blank with the correct word or words.

11. _____ and _____ are two body hormones whose structures differ by only one methyl group.

12. _____ is an alkaloid found in cigarettes. In pure form it is _____ (mildly, extremely) toxic, and acts on the body by _____ the central nervous system.

13. _____ is a stimulant found in coffee, tea, and cocoa.

14. _____ and _____ are extracted from opium poppies, and have a _____ effect on humans. _____ is used to reduce pain after a serious accident or surgery. _____ is found in some cough medicines.

15. _____ is a synthetic alkaloid produced from morphine.

16. _____ is the name given to the natural substances produced by the brain that produce effects similar to the opiates.

17. The barbiturates act by _____ the central nervous system and are used in such products as _____ and _____ .

18. _____, a thiobarbiturate, is a very effective, fast-acting anesthetic.

Answers to Self-Test Questions in Chapter 16

1. (a) $CH_3(CH_2)_4CH_2NH_2$ (b) $CH_3CH_2CH_2CH_2-NH-CH_2CH_2CH_2CH_3$

(c) $CH_3CH_2-\underset{\underset{CH_3}{|}}{N}-CH_2CH_2CH_3$ (d) $CH_3-\underset{\underset{CH_3-N-H}{|}}{CH}-CH_2OH$

2. (a) primary (b) secondary (c) tertiary (d) secondary
3. (a) ammonia (b) diphenylamine (c) butylamine
(d) ethyldipropylamine **4.** b,a,c,d

5. (a) $CH_3CH_2CH_2NH_3^+Cl^-$ (b) $CH_3CH_2\underset{\underset{CH_3}{|}}{N}H_2^+OH^-$ (c) $(CH_3CH_2)_4N^+Br^-$

6. c,d,f **7.** (c) 2-methylbutanamide (d) N-ethylpropanamide
(f) N-isopropyl-N-methylacetamide

8. (a) $\langle O \rangle -\overset{\overset{O}{\|}}{C}NH_2$ (b) $CH_3(CH_2)_6\overset{\overset{O}{\|}}{C}NHCH_2CH_3$ (c) $CH_3(CH_2)_4\overset{\overset{O}{\|}}{C}N\underset{\underset{CH_3}{|}}{CH_3}CH_2$

9. (a) $\langle O \rangle -NH_2 + CH_3CH_2\overset{\overset{O}{\|}}{C}OH$ (b) $CH_3CH_2C\equiv N + H_2O$

(c) $CH_3CH_2\overset{\overset{O}{\|}}{C}OH + NH_3$ (d) $(CH_3CH_2)_2NH + CH_3CH_2CH_2\overset{\overset{O}{\|}}{C}OH$

10. The nitrogen on acetamide has two hydrogen atoms attached to it that can form hydrogen bonds with oxygens on other acetamide molecules. N,N-dimethylacetamide has no hydrogen atoms attached to the nitrogen and, therefore, cannot form hydrogen bonds.
11. epinephrine, norepinephrine **12.** nicotine, extremely, stimulating **13.** caffeine **14.** morphine, codeine, narcotic, morphine, codeine **15.** heroin **16.** endorphins **17.** depressing, sleeping pills, sedatives (or anesthetics or hypnotics)
18. sodium pentothal

Chapter 17   CARBOHYDRATES

In this chapter we begin studying the compounds of life by discussing the large class of compounds called carbohydrates. Compounds in this class include sugars, starches, and cellulose. Carbohydrates are found mainly in plant material; they function both as the supporting structure and food storage molecules of the plant. Carbohydrates can be classified by the size of the molecule.

17.1 Classification

Monosaccharides are simple sugars that cannot be broken into smaller units upon hydrolysis. Disaccharides contain two monosaccharides; they will break apart to form two simple sugars upon hydrolysis. Polysaccharides are polymers containing many monosaccharide units.

Monosaccharides may be further classified by the number of carbon atoms in the molecule and by the carbonyl functional group found in the molecule.

Important Terms

monosaccharide	disaccharide	polysaccharide
aldose	ketose	

Self-Test

For questions 1 to 7, fill in the blank with the correct word or words.

1. _____ is the name given to the large class of compounds that includes sugars, starches, and gums.

2. These compounds are found mainly in _____, and their functions are _____ and _____ .

3. _____ yield three or more simple sugars upon hydrolysis.

4. _____ are carbohydrates that cannot be broken down by hydrolysis.

5. _____ yield two simple sugars upon hydrolysis.

6. A monosaccharide that contains five carbon atoms and a ketone functional group is a _____ .

7. A monosaccharide that contains six carbon atoms and an aldehyde functional group is an _____ .

17.2 Glucose

Simple sugars are white crystalline solids that are soluble in water and have a sweet taste. The most abundant monosaccharides are the hexoses, and the most abundant hexose is glucose. Glucose is an aldohexose. It is the immediate energy source for cells in living organisms and is the monomer unit for many important polysaccharides.

Glucose is found in three forms in water solution. Two different ring forms (alpha and beta) and a straight chain form exist in equilibrium with one another, with the ring forms being much more common. We can write the structure of glucose in straight chain form, or in ring form using Haworth projections. The rings result from the formation of an internal hemiacetal between the aldehyde group on carbon number 1 and the alcohol group on carbon number 5. The formation of a hemiketal between the ketone group and an alcohol group will result in similar ring forms in the ketoses.

Important Terms

hexose	aldohexose	glucose
hemiacetal	hemiketal	dextrose
Haworth projection	alpha ring form	beta ring form

Self-Test _____

For questions 8 to 10, fill in the blank with the correct word or words.

8. The monosaccharide _____ is also called dextrose.

9. The major function of glucose in living cells is _____ .

10. The three structural forms of glucose in water solution are _____ , _____ , and _____ .

11. Which of the following structures is:
    (a) the alpha form of glucose   (b) the beta form of glucose

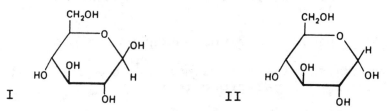

17.3 Optical Isomerism (optional section)

A carbon atom that has four different groups attached to it is called a chiral carbon, and will possess a property called chirality. Such chiral carbons have a mirror image that is not superimposable on itself. Molecules that have a chiral carbon and are mirror images will have identical physical properties except one: they will interact differently with polarized light. These mirror image molecules are optical isomers that will rotate the plane of polarized light in opposite directions.

Carbohydrate molecules possess chiral carbons, so they are optically active. The D and L classification system of optical isomers is based on the three-carbon compound glyceraldehyde. Most naturally occurring carbohydrates belong to the D family of isomers.

Important Terms

optical isomerism	optical isomer	chirality
chiral center	chiral molecule	

Self-Test _____

For questions 12 to 19, answer true (T) or false (F).

12. When bonded to four different groups, a carbon atom possesses a property called chirality.

13. Carbon #2 in 2-butanol is a chiral center.

$$CH_3 - \underset{\underset{H}{|}}{\overset{\overset{OH}{|}}{C}} - CH_2CH_3 \qquad (\text{2-butanol})$$

14. A chiral molecule can be superimposed on its mirror image.

15. Polarized light is light that vibrates in several planes.

16. Molecules that are nonsuperimposable mirror images have the same physical and chemical properties.

17. Optical isomers will rotate the plane of polarized light in a clockwise direction.

18. If a chiral molecule rotates the plane of polarized light in a clockwise direction, its mirror image will rotate the plane of polarized light in a counterclockwise direction.

17.4 Fructose

Fructose is a ketohexose that, like glucose, exists in alpha and beta ring forms. When combined with glucose, it forms the disaccharide sucrose. Fructose is found in fruit juices and honey.

Important Terms

    fructose   ketohexose   hemiketal

17.5 Galactose

Galactose is an aldohexose that is not found in nature as a free monosaccharide. Rather, it is a component of larger molecules such as lactose, glycolipids, and agar-agar.

Important Terms

    galactose   lactose

17.6 Pentoses

Pentoses are five-carbon sugars. Important pentose sugars are arabinose, xylose, ribose, and deoxyribose.

Important Term

    pentose   arabinose   xylose
    ribose   deoxyribose

17.7 Reducing Sugars

Carbohydrates that have a free aldehyde or ketone group are called reducing sugars because they will reduce basic solutions of mild oxidizing agents such as Cu^{2+} or Ag^+. Benedict's test is very useful for determining the presence of a reducing sugar. A reducing sugar will reduce the Cu^{2+} ions in the Benedict's test solution to Cu^+ ions, resulting in the formation of a brick-red precipitate of Cu_2O. Clinitest tablets make use of this reaction to test for sugar in the urine. The Tollens' pentose test can be used to determine if the sugar found using the Benedict's test is a hexose or pentose.

Important Terms

    reducing sugar   Benedict's test   Tollens' pentose test

For questions 19 to 35, select the best answer or answers from the choices given.

19. Which of the following is an aldohexose?
    (a) ribose   (c) fructose
    (b) glucose   (d) arabinose

20. Carbohydrates that contain two simple sugars are called
    (a) monosaccharides   (c) disaccharides
    (b) polysaccharides   (d) glycolipids

21. The sugar sucrose contains the simple sugars
    (a) ribose   (c) fructose
    (b) glucose   (d) galactose

22. One of the simple sugars found in lactose is
    (a) ribose   (c) fructose
    (b) galactose   (d) xylose

23. The solution you would use to identify the presence of a reducing sugar is
    (a) Tollens'   (c) Benedict's
    (b) Clinitest   (d) Fehling's

24. The solution you would use to distinguish between a hexose and pentose sugar is
    (a) Tollens'   (c) Benedict's
    (b) Clinitest   (d) Fehling's

25. Which of the following is a ketohexose?
    (a) ribose   (c) fructose
    (b) glucose   (d) galactose

26. The name given to the ring structure of glucose when the hydroxyl group on carbon #1 is above the ring is
    (a) alpha form   (c) beta form
    (c) hemiacetal   (d) hemiketal

27. The monomer sugar found in agar-agar is
    (a) ribose   (c) fructose
    (b) glucose   (d) galactose

28. A carbohydrate that has a free aldehyde or ketone group is called a
    (a) reducing sugar   (c) disaccharide
    (b) polysaccharide   (d) hemiacetal

29. Which of the following is an aldopentose?
    (a) ribulose   (c) ribose
    (b) galactose   (d) fructose

30. The name given to the ring structure of glucose when the hydroxyl group on carbon #1 is below the ring is
    (a) alpha form   (c) beta form
    (c) hemiacetal   (d) hemiketal

31. The five-member ring structure of fructose is called a(n)
    (a) acetal linkage   (c) ketohexose
    (b) hemiacetal   (d) hemiketal

32. Another name for blood sugar is
    (a) glucose   (c) ribose
    (b) galactose   (d) fructose

33. Carbohydrates that contain many simple sugars are called
    (a) monosaccharides   (c) disaccharides
    (b) polysaccharides   (d) glycolipids

34. The five-member ring structure of glucose is called a(n)
    (a) acetal linkage   (c) aldohexose
    (b) hemiacetal   (d) hemiketal

35. A carbohydrate that cannot be broken into smaller units upon hydrolysis is called a
    (a) monosaccharide   (c) disaccharide
    (b) polysaccharide   (d) glycolipid

17.8 Maltose

Maltose (or malt sugar) is a disaccharide containing two units of glucose connected by an α,1:4 linkage. This bond is a glycosidic (or acetal) linkage formed between carbon number one on one glucose molecule in the alpha form and carbon number four on the second glucose molecule. The aldehyde group on the second glucose molecule is not involved in the acetal linkage; therefore, maltose is a reducing sugar.

Important Terms

    maltose   acetal linkage   condensation reaction
    glycosidic linkage α,1:4 linkage

17.9 Lactose

Lactose is a disaccharide containing a glucose unit and a galactose unit. The acetal linkage is between carbon 1 of galactose in the beta form and carbon 4 of glucose, and is called a β,1:4 linkage. Lactose is found in the milk of mammals. The molecule contains a free aldehyde group and is a reducing sugar.

Important Term

    lactose β,1:4 linkage

17.10 Sucrose

Sucrose (or table sugar) is a disaccharide containing a unit of glucose and a unit of fructose. It is the sugar we use in cooking. The linkage in the sucrose molecule is an α,1:2 linkage between the aldehyde group on the glucose and the ketone group on the fructose. Therefore, sucrose is not a reducing sugar.

Important Terms

    sucrose α,1:2 linkage

Self-Test _____

For questions 36 to 41, fill in the blank with the correct word or words.

36. An _____ linkage is formed in a condensation reaction between a hemiacetal and an alcohol.

37. This linkage is _____ (more, less) stable than a hemiacetal linkage.

38. _____ is composed of glucose and fructose.

39. Maltose contains two _____ units held together by a _____ linkage.

40. _____ contains a unit of galactose and a unit of glucose, and is found in _____ .

41. When produced in the controlled germination of grains, _____ (also called malt) is used in the production of _____ .

17.11 Starch

Starch is a polymer of glucose used by plants as the storage form for energy. Starch is a mixture of two polysaccharides: amylose and amylopectin. Amylose is a linear polysaccharide in which all the glucose units are connected by α,1:4 linkages. Amylopectin is a branched polysaccharide containing glucose units connected by α,1:4 linkages in the nonbranching portions, and α,1:6 linkages at the branch points.

Dextrins are polysaccharides formed by the partial hydrolysis of starch. They are formed in the baking bread, and are used as adhesives on stamps and envelopes.

Important Terms

    starch   amylose   amylopectin
    α,1:6 linkage   dextrins

17.12 Glycogen

Glycogen is a highly branched polymer of glucose that is the storage form of glucose in animals. As in amylopectin, the branching is a result of α,1:6 linkages between the glucose units. The level of glucose in the blood remains relatively constant because the body stores excess glucose as glycogen in the liver right after a meal and then slowly releases the glucose back into the blood as the level falls.

Important Term

    glycogen

17.13 Cellulose

Cellulose is a polymer of glucose produced by plants to form the supportive framework of the plant. The rigidity that cellulose gives to plants is a result of hydrogen bonding between cellulose molecules. The glucose units in cellulose are held together by β,1:4 linkages. We do not have enzymes in our digestive tract that can break apart β,1:4 linkages between glucose molecules. Therefore, any cellulose that we eat passes through our digestive tract undigested.

Important Terms

    cellulose β,1:4 linkage

17.14 Dextran

Dextran is a glucose polymer produced by bacteria. When it is partially hydrolyzed, dextran can be used as a plasma substitute in the treatment of shock. Dextrans, formed by bacteria in the mouth, are the "glue" that holds plaque on the teeth.

Important Terms

    dextran   plaque

17.15   Iodine Test

The iodine test is used to detect the presence of starch. It will turn an intense blue-black color when starch is present.

Important Term

iodine test

Self-Test _____

For questions 42 to 59, match the statement with the correct answer or answers from column B. An item in column B may be used more than once.

Column B

42. Hemiacetal and an alcohol

43. Contains galactose and glucose

44. Storage form of glucose in plants

45. Contains α,1:6 linkages

46. Forms support structures in plants

47. Can be used as a blood plasma substitute

48. Identifies the presence of starch

49. Used in the manufacture of beer

50. Found in the milk of mammals

51. Contains glucose and fructose

52. Storage form of glucose in animals

53. Forms golden brown color of bread crusts

54. Contains β,1:4 linkages between glucose molecules

55. Contains two glucose units

56. Table sugar

57. Polysaccharides forming starch

Column B
A. acetal
B. maltose
C. lactose
D. sucrose
E. starch
F. glycogen
G. amylose
H. amylopectin
I. dextrin
J. cellulose
K. dextran
L. iodine test

58. Used as an adhesive

59. Used to produce plastics, films, and guncotton

17.16 Photosynthesis

Photosynthesis is the process by which plants capture the sun's energy and use it to produce carbohydrates. The process is not entirely understood, but it involves two series of reactions: the light and dark reactions. The light reactions require sunlight and molecules of chlorophyll, and produce oxygen and energy-rich molecules. The dark reactions then use the energy-rich molecules to form glucose and other organic molecules from carbon dioxide. The overall reaction for photosynthesis is

$$6CO_2 \; + \; 6H_2O \; \xrightarrow[\text{chlorophyll}]{\text{light}} \; C_6H_{12}O_6 \; + \; 6O_2$$

Important Terms

photosynthesis   chloroplast   chlorophyll

Self-Test _____

For questions 60 to 64, fill in the blank with the correct word or words.

60. _____ is the process by which plants produce carbohydrates.

61. This process involves two series of reactions called _____ and _____ reactions.

62. The _____ reactions require the presence of chlorophyll and sunlight, and produce _____ and _____ .

63. The _____ reactions, which can occur without sunlight, produce _____ from _____ and _____ .

64. All of these reactions take place in the _____ of the plant cell.

Answers to Self-Test Questions in Chapter 17

1. carbohydrate 2. plants, support, energy storage
3. polysaccharides 4. monosaccharides 5. disaccharides
6. ketopentose 7. aldohexose 8. glucose 9. supply immediate
energy needs 10. alpha ring, beta ring, straight chain
11. (a) II (b) I 12. T 13. T 14. F 15. F 16. F 17. F
18. T 19. b 20. c 21. b,c 22. b 23. b,c,d 24. a 25. c
26. c 27. d 28. a 29. c 30. a 31. d 32. a 33. b 34. b
35. a 36. acetal or glycosidic 37. more 38. sucrose
39. glucose, α,1:4 40. lactose, milk of mammals 41. maltose,
beer 42. A 43. C 44. E 45. F,H 46. J 47. K 48. L 49. B
50. C 51. D 52. F 53. I 54. J 55. B 56. D 57. G,H 58. I
59. J 60. photosynthesis 61. light, dark 62. light, oxygen,
energy-rich molecules 63. dark, glucose, carbon dioxide,
energy-rich molecules 64. chloroplasts

Chapter 18   LIPIDS

We now turn to a second major class of the compounds of life: the lipids. In the last chapter, we saw that carbohydrates are highly polar substances, and that monosaccharides and disaccharides are soluble in water. Lipids, by contrast, are oily or waxy substances that are not soluble in water. As a result they are found in the structural components of cells. A second major function of lipids is to store energy in cells.

18.1 What are Lipids?

Lipids are oily or waxy substances that function as energy storage molecules and as structural components of cells. They can be categorized as saponifiable (those that can be hydrolyzed by a base) and nonsaponifiable. The saponifiable lipids are the simple lipids that yield fatty acids and alcohol upon hydrolysis. The compound lipids, when hydrolyzed, yield fatty acids, alcohol, and other compounds.

Important Terms

      simple lipid   compound lipid   saponifiable

18.2 Fats and Oils

The neutral fats, also called triacylglycerols or triglycerides, are esters of glycerol and three fatty acids. They function as energy storage molecules, and include both simple and mixed triacylglycerols.

Important Terms

      neutral fat   simple triacylglycerol
      triglyceride   mixed triacylglycerol
      triacylglycerol

18.3 Fatty Acids

Fatty acids are long-carbon-chain carboxylic acids that are formed upon hydrolysis of triacylglycerols. Saturated fatty acids contain only carbon-to-carbon single bonds and are solids at room temperature. Unsaturated fatty acids have one or more carbon-to-carbon double bonds and are liquids at room temperature. Fats and oils differ in the number of unsaturated fatty acids they contain. We can indicate the position and number of double bonds in an unsaturated fatty acid as follows: linoleic acid (18:2n-6). The (18:2n-6) means that linoleic acid has 18 carbons and 2 double bonds, the first of which is on the sixth carbon from the methyl end of the molecule.

$$CH_3CH_2CH_2CH_2CH_2CH=CHCH_2CH=CHCH_2CH_2CH_2CH_2CH_2COOH$$

Important Terms

fatty acid	unsaturated fatty acid
oil	saturated fatty acid
fat	

18.4 Essential Fatty Acids

Our bodies cannot synthesize adequate amounts of certain fatty acids that are known as essential fatty acids. These fatty acids are required to maintain normal cell membranes and skin. As a result, it is important that our diets contain an adequate supply of linoleic and arachidonic acids. The body uses arachidonic acid as the starting material to synthesize a very important class of compounds called eicosanoids.

Important Terms

essential fatty acid   eicosanoid

Self-Test _____

For questions 1 to 16, choose the correct answer or answers.

1. Another name for triacylglycerol is
   (a) triglyceride   (c) simple lipid
   (b) neutral fat   (d) compound lipid

2. A fatty acid that contains only carbon-to-carbon single bonds is called
   (a) saponifiable   (c) nonsaponifiable
   (b) saturated   (d) unsaturated

3. The essential fatty acids are
   (a) palmitic   (c) arachidonic
   (b) linoleic   (d) stearic

4. The hydrolysis of a triacylglycerol produces
   (a) fatty acids   (c) simple lipids
   (b) compound lipids   (d) glycerol

5. This compound contains a large number of unsaturated fatty acids.
   (a) compound lipid   (c) oil
   (b) fat   (d) simple triacylglycerol

6. A lipid that can be hydrolyzed by a base is
   (a) saponifiable   (c) nonsaponifiable
   (b) saturated   (d) unsaturated

7. This compound yields fatty acids, alcohol, and other compounds upon hydrolysis.
   (a) simple lipid   (c) simple triacylglycerol
   (b) compound lipid   (d) mixed triacylglycerol

8. A neutral fat that contains different fatty acids is called a
   (a) simple lipid   (c) simple triacylglycerol
   (b) compound lipid   (d) mixed triacylglycerol

9. One of the most abundant fatty acids is
   (a) butyric acid   (c) arachidic acid
   (b) linolenic acid   (d) oleic acid

10. The tissue that contains fat cells is called
    (a) adipose tissue   (c) epithelial tissue
    (b) connective tissue   (d) muscle tissue

11. The body requires these in the diet
    (a) triacylglycerols   (c) essential fatty acids
    (b) compound lipids   (d) saturated fatty acids

12. Compounds that contain only fatty acids and alcohol.
    (a) simple lipid   (c) simple triacylglycerol
    (b) compound lipid   (d) mixed triacylglycerol

13. A fat that has three identical fatty acids is called a
    (a) simple lipid   (c) simple triacylglycerol
    (b) compound lipid   (d) mixed triacylglycerol

14. A fatty acid that contains carbon-to-carbon double bonds is
    (a) saponifiable   (c) nonsaponifiable
    (b) saturated   (d) unsaturated

15. This compound is used by the body to synthesize prostaglandins.
    (a) palmitic acid   (c) arachidonic acid
    (b) oleic acid   (d) stearic acid

16. This lipid contains more saturated than unsaturated fatty acids.
    (a) compound lipid   (c) oil
    (b) fat   (d) simple triacylglycerol

18.5 Waxes

Waxes are esters of long-chain fatty acids and long-chain
alcohols. They are insoluble in water, flexible, and
nonreactive. They form various types of protective coatings on
plants and animals.

Important Term

   wax

18.6 Iodine Number

The iodine number measures the degree of unsaturation of a
triacylglycerol. The higher the iodine number, the greater the
unsaturation. Fats have iodine numbers below 70, and oils above
70.

Important Term

   iodine number

18.7 Hydrogenation

Hydrogenation is the process of adding hydrogen to the double
bonds in the fatty acids of an oil, converting it to a solid fat.
This process is used commercially to produce solid shortenings
and margarine.

18.8 Formation of Acrolein

When heated to high temperatures, fats will hydrolyze. The
resulting glycerol reacts to form acrolein, a substance with an
easily detected, unpleasant odor.

18.9 Rancidity

A fat becomes rancid when it undergoes hydrolysis or oxidation.
Butter fat can be hydrolyzed by microorganisms in the air,
releasing the strong-smelling butyric acid. Oxygen in the air
can oxidize unsaturated fats and oils to form short-chain fatty
acids and aldehydes having disagreeable odors. This oxidation
can be slowed by adding antioxidants to the food product.

Important Terms

   rancidity   antioxidant

18.10 Hydrolysis

Hydrolysis of triacylglycerols occurs in the presence of steam, hot mineral acids, or enzymes. The hydrolysis yields three fatty acids and glycerol.

Important Term

hydrolysis

18.11 Saponification

Hydrolysis of triacylglycerols in the presence of a strong base yields glycerol and three fatty acid salts. Such salts are soaps. The type of metal ion in the salt and the amount of unsaturation in the fatty acid will determine the nature of the soap. Calcium, magnesium, and iron soaps are insoluble in water. These ions are often found in hard water and will cause the soap to precipitate out, forming soap scum and decreasing the cleansing action of the soap. Synthetic detergents are superior to soaps because their calcium and magnesium salts are soluble in water, and they are not affected by pH.

A soap molecule contains a nonpolar tail that can dissolve in grease and oil, and a polar head that is soluble in water. Soap acts by emulsifying the grease, making it form small colloidal particles that can be washed away by water.

Important Terms

saponification   soap   detergent

Self-Test _____

For questions 17 to 25, fill in the blanks with the correct word or words.

17. When fats are heated to high temperatures, the odor of _____ can be detected. It is formed from _____ released when the fats hydrolyze.

18. The iodine test measures the _____ of a triacylglycerol.

19. _____ have iodine numbers greater than 70, and _____ have iodine numbers less than 70.

20. _____ is the process by which hydrogen is added to the double bonds in a triacylglycerol, forming a _____ fat from a _____ fat.

21. If arachidonic acid were treated by the process in question 20, _____ acid would be produced.

22. When a fat or oil undergoes oxidation, it may develop a disagreeable _____ or _____ , and is called _____ .

23. Hydrolysis of a simple triacylglycerol yields _____ and _____ .

24. _____ is the hydrolysis of a triacylglycerol in the presence of a strong base; it produces _____ and _____ , which are called soaps.

25. Liquid soaps are produced when the hydrolysis is carried out in the presence of _____ ; solid soaps result when the hydrolysis is carried out in the presence of _____ .

For questions 26 to 33, answer true (T) or false (F).

26. Soap can emulsify grease.

27. Magnesium soaps are soluble in water, but calcium soaps are not.

28. The nonpolar tail of the soap molecule dissolves in the water, and the polar head of the molecule dissolves in the grease.

29. Hard water contains metal ions that form insoluble salts with soap.

30. Water softeners work by removing metal ions from the water and replacing them with ions that don't interfere with the soap.

31. Changing the pH of the water will affect the cleansing action of detergents.

32. The calcium salts of detergents are water soluble.

33. Synthetic detergents have a molecular shape similar to that of fatty acid soaps.

18.12 Glycerol-Based Phospholipids: Phosphoglycerides

The phosphoglycerides are compound lipids containing glycerol, two fatty acids, phosphoric acid, and a nitrogen compound. The structure of these compounds is similar to the soaps, and they have good emulsifying properties. They also form the double-layer framework for cell membranes.

Phosphatidylcholine (PC or lecithin) is a phosphoglyceride in which the nitrogen compound is choline. PC is important in the metabolism of fats in the liver, is a source of phosphate for tissue formation, and helps in the transport of fats. A derivative of PC, dipalmitoyl phosphatidylcholine (DPPC) acts as a surfactant in the lungs to reduce the surface tension of water and keep the alveoli from collapsing.

Phosphatidylethanolamines (cephalins) are phosphoglycerides in which the nitrogen compound is ethanolamine. They are important in the clotting mechanism of the blood, and also serve as a source of phosphate for the formation of new tissue.

Important Terms

phospholipid   phosphoglyceride
phosphatidylcholine   lecithin
phosphatidylethanolamine   cephalins
respiratory distress syndrome

18.13 Sphingosine-Based Phospholipids: Sphingolipids

Sphingolipids are phospholipids that contain the alcohol sphingosine instead of glycerol. The most common sphingolipid is sphingomyelin, which forms part of the protective coating (called the myelin sheath) around nerve cells.

Important Terms

sphingolipid   sphingosine
myelin sheath   sphingomyelin

18.14 Glycolipids

Glycolipids have a structure very similar to phospholipids, except that the phosphate group is replaced by a sugar that is usually galactose, but may also be glucose. Cerebrosides are glycolipids containing sphingosine, and are found in high concentrations in the myelin sheath.

Important Terms

glycolipid   cerebroside

Self-Test _____

For questions 34 to 44, match the statement with the correct answer or answers in column B. An item in column B may be used more than once.

34. Forms the framework of cell membrane

35. Contains the alcohol glycerol

36. Ester of a long-chain alcohol and a long-chain fatty acid

37. Contains the alcohol sphingosine

38. Important emulsifying agent

39. Similar to phospholipids but contains a sugar group rather than a phosphate group

40. Source of phosphate for tissue formation

41. Component of protective coating of skin, feathers, and leaves

42. Protective coating around nerve cells

43. Found in blood platelets

44. Forms part of the myelin sheath

Column B
A. wax
B. phosphoglycerides
C. phosphatidylcholine
D. sphingolipids
E. myelin sheath
F. glycolipids
G. phosphatidyl-
   ethanolamine

18.15 Steroids

Steroids are a large class of compounds having a wide range of functions (see Table 18.5 in the text). The basic structure of all steroids is a nucleus containing three six-carbon rings and one five-carbon ring.

Important Term

steroid

18.16 Cholesterol

Sterols are steroid alcohols. Cholesterol is the most abundant sterol and is found in most animal tissue. Cholesterol is synthesized by the liver from acetyl-CoA, is used in the synthesis of bile salts, sex hormones, and vitamin D, and plays a role in the development of atherosclerosis.

Important Terms

sterol   cholesterol

18.17　Cellular Membranes

Cellular membranes contain lipids and proteins and small amounts of carbohydrate. The membrane structure is a lipid bilayer containing two rows of phospholipids with their polar heads to the outside and their nonpolar tails to the inside. The protein molecules are found spaced throughout the membrane (See Figure 18.8 in the text).

Self-Test _____

For questions 45 to 52, fill in the blank with the correct word or words.

45.　Vitamin D, cortisone, and progesterone are all members of the class of compounds called _____ .

46.　Steroids belong to the large class of lipids that are _____ (not hydrolyzed by base).

47.　Steroids all share the same common _____ .

48.　Cortisone and testosterone are steroid _____ .

49.　Digitoxigenin is used as a drug to treat _____ .

50.　Sterols are steroids containing a(n) _____ group. The most abundant sterol is _____ . This sterol is synthesized in the _____ from _____ . It is also absorbed from foods such as _____ and _____ . It is carried in the blood in the form of _____ , _____ , and _____ .

51.　The two major components of cellular membranes are _____ and _____ .

52.　Describe the structure of a cellular membrane.

_____

1. a,b **2.** b **3.** b,c **4.** a,d **5.** c **6.** a **7.** b **8.** d **9.** d **10.** a
11. c **12.** a,c,d **13.** c **14.** d **15.** c **16.** b **17.** acrolein,
glycerol **18.** unsaturation **19.** oils, fats **20.** hydrogenation,
solid (saturated), liquid (unsaturated) **21.** arachidic **22.** odor,
taste, rancid **23.** glycerol and three molecules of the same fatty
acid **24.** saponification, glycerol, fatty acid salts
25. potassium hydroxide, sodium hydroxide **26.** T **27.** F **28.** F
29. T **30.** T **31.** F **32.** T **33.** T **34.** B,D **35.** B,C,G **36.** A
37. D **38.** C **39.** F **40.** C,G **41.** A **42.** E **43.** G **44.** D
45. steroids **46.** nonsaponifiable **47.** four ring structure
48. hormones **49.** heart disease **50.** hydroxyl (alcohol),
cholesterol, liver, acetyl-CoA, chylomicrons, LDL, HDL
51. lipids, proteins **52.** The cell membrane is a lipid bilayer
with proteins and carbohydrate components imbedded in the
bilayer.

Chapter 19   PROTEINS

This chapter discusses the most complex and varied molecules of
life, the proteins. First we present several methods of
classifying this large group of compounds. Then, we discuss the
properties of amino acids, the monomers that form the polymer
proteins. Proteins have complex structures, and their structure
determines their functions. In the second section of this
chapter, we discuss protein structure and the factors that affect
this structure.

19.1   Classification of Proteins

Proteins are extremely important biological molecules, serving
many functions in the cell. They are very large polymers of
amino acids; their size is in the range of colloidal particles.

Proteins may be classified as simple proteins containing only
amino acids, or conjugated proteins containing other inorganic or
organic components called prosthetic groups. Globular proteins
are soluble in water, are quite sensitive to their chemical
environment, and often function as enzymes. Fibrous proteins are
insoluble in water, are fairly insensitive to their chemical
environment, and have a structural or protective function.
Proteins can also be categorized on the basis of their functions
in living organisms. Table 19.2 in the text lists some of these
categories.

Important Terms

         amino acid   simple protein
         globular protein   conjugated protein
         fibrous protein   prosthetic group

19.2   Amino Acids

Amino acids contain both an amino and a carboxyl functional
group. Most naturally occurring proteins are composed of
combinations of only 20 different amino acids, which can be
identified by their R-group side chains.

Self-Test _____

For questions 1 to 10, choose the best answer or answers.

1. Because of their size, proteins belong to which class of compounds?
   (a) crystalloids   (c) glycoproteins
   (b) colloids   (d) lipoproteins

2. Components of proteins that are other than amino acids are called
   (a) conjugated groups   (c) prosthetic groups
   (b) metal ion   (d) enzymes

3. Proteins that function as structural components of organisms are called
   (a) simple proteins   (c) globular proteins
   (b) conjugated proteins   (d) fibrous proteins

4. The monomer units of proteins are
   (a) sugars   (c) enzymes
   (b) amino acids   (d) carboxylic acids

5. These proteins are soluble in water.
   (a) simple proteins   (c) globular proteins
   (b) conjugated proteins   (d) fibrous proteins

6. These proteins are physically tough.
   (a) glycoproteins   (c) globular proteins
   (b) conjugated proteins   (d) fibrous proteins

7. These proteins function as enzymes.
   (a) glycoproteins   (c) globular proteins
   (b) contractile proteins   (d) fibrous proteins

8. Proteins that contain only amino acids are called
   (a) simple proteins   (c) globular proteins
   (b) conjugated proteins   (d) fibrous proteins

9. Hemoglobin belongs to this class of proteins.
   (a) structural protein   (c) storage protein
   (b) transport protein   (d) protective protein

10. Antibodies belong to this class of proteins.
    (a) structural protein   (c) storage protein
    (b) contractile protein   (d) protective protein

19.3 The L-Family of Amino Acids (optional section)

With the exception of glycine, all amino acids that are found in proteins are optically active and belong to the L-family.

11. Draw the structure for (a) L-valine and (b) L-serine.

19.4 Essential Amino Acids

Ten of the 20 amino acids found in proteins cannot be synthesized in sufficient amounts by the human body. These amino acids, called the essential amino acids, must be supplied in the diet. Protein that contains all 10 essential amino acids is called adequate protein. Animal protein and milk are adequate proteins, but many vegetable proteins are not.

Important Terms

essential amino acid   adequate protein

19.5 Acid-Base Properties

When dissolved in water, amino acids exist as zwitterions, or dipolar ions. Amino acids are amphoteric: they can act either as an acid or a base. Proteins function as buffers in the blood.

Important Terms

amphoteric   zwitterion

19.6 Isoelectric Point

The isoelectric point is the pH at which an amino acid or protein is electrically neutral and will not migrate in an electric field. Each amino acid and protein has a characteristic isoelectric point. Proteins are the least soluble at their isoelectric points.

Important Term

isoelectric point (pI)

For questions 12 to 16, fill in the blank with the correct word or words.

12. Amino acids are amphoteric, which means that they _____.

13. In water, amino acids exist as highly polar ions called
    _____ .

14. Amino acids that the body cannot synthesize in sufficient
    amounts are called _____ . The protein in beef is an
    example of an _____ protein, but the protein in rice is
    not because it doesn't contain sufficient amounts of _____
    or _____ .

15. The pH at which an amino acid or protein will not migrate in
    an electric field is called its _____ . At a pH more
    basic than this point, a protein will carry a net _____
    charge and will migrate toward the _____ pole in an
    electric field.

16. Proteins are the least soluble at a pH _____ (less than,
    equal to, or more than) their isoelectric points.

17. (a) Draw the structure of alanine in neutral water.
    (b) Show how alanine can act as a buffer and can protect
        against changes in pH when an acid or a base is added
        to the solution.

For questions 18-20 describe (a) the charge [positive, negative,
no charge] the molecule will carry at a pH of 6, and (b) toward
what pole [negative or positive] it will migrate if placed in an
electric field at pH of 6.

18. alanine   19. egg albumin   20. hemoglobin

19.7 Primary Structure

The primary structure of a protein is its sequence of amino
acids. The amino acids are linked by peptide bonds, which are
amide linkages formed by a condensation reaction between the
amino acids. The peptide bond is stable, and can be broken only
by certain enzymes or by acid or base hydrolysis. A dipeptide
contains two amino acids, a tripeptide contains three amino
acids, an oligopeptide four to ten amino acids, and a polypeptide
more than ten.

The amino acid sequence of a protein can be indicated using
three-letter abbreviations for the names of the amino acids.
Dashes or dots are used to show the peptide bonds. The end of
the molecule containing the free amino group is called the
N-terminal end, and is usually written first in the sequence.
The end containing the free carboxyl group is called the
C-terminal end, and is written last in the sequence.

The amino acid sequence is critical to the structure and function
of the protein. A change in just one amino acid can totally
disrupt the normal functioning of the protein.

primary structure   dipeptide
peptide bond   tripeptide
N-terminal end   oligopeptide
C-terminal end   polypeptide

Self-Test

21. (a) Draw the structure of the tripeptide formed between alanine, serine, and valine. Put valine on the N-terminal end and serine on the C-terminal end and circle the peptide bonds.
    (b) Write the three letter abbreviation for this sequence.

19.8 Secondary Structure

The secondary structure of proteins is the shape of specific regions of the polypeptide backbone. It results from hydrogen bonding between the amino acids, and takes on several configurations: the alpha helix, the beta configuration or pleated sheet, and the triple helix.

In the alpha helix the amino acids form loops, with the R-groups on the amino acids extending to the outside of the helix. In the beta configuration several polypeptide chains are lined up next to one another, held together by hydrogen bonds. The R-groups on the amino acids extend above and below the plane of the sheet. The triple helix consists of three polypeptide chains twisted around each other to form a helix.

Each protein will have its own characteristic secondary structure determined by its sequence of amino acids. Some large globular proteins may have regions of alpha helix and regions of beta configuration in their structure. Unlike peptide bonds, hydrogen bonds are easily disrupted by changes in pH, temperature, solvents, or salt concentrations.

Important Terms

secondary structure   alpha helix
beta pleated sheet   keratin
beta configuration   collagen
triple helix   silks

19.9 Tertiary Structure

Globular proteins have a tightly folded three-dimensional structure that results from disulfide bridges, hydrogen bonding, salt bridges, and hydrophobic interactions. This shape is the native state or native configuration of the protein molecule.

Disulfide bridges form between the side chains of two cysteine molecules, forming a covalent linkage between two regions on one polypeptide chain or between two polypeptide chains. These linkages are not sensitive to changes in pH, salt concentration, solvents, or temperature, and can be broken only by reduction.

Salt bridges are ionic attractions between the charged side chains of amino acids such as lysine or aspartic acid, and can be disrupted by changes in pH.

The hydrophobic side chains of certain amino acids tend to congregate toward the inside of the molecule, away from the polar water molecules. These hydrophobic interactions are key to the shape of the protein molecule.

Important Terms

> tertiary structure   native state
> native configuration   disulfide bridge
> hydrophobic interaction   salt bridge
> hydrogen bonding

19.10 Quaternary Structure

The quaternary structure of a protein is the way in which two or more polypeptide chains are held together in a three-dimensional arrangement. Hydrogen bonding, salt bridges, and hydrophobic interactions may all be involved in holding the chains together.

Important Term

> quaternary structure

Self-Test _____

For questions 22 to 31, fill in the blank with the correct word or words.

22. The _____ structure of a protein is the sequence of amino acids in the protein. These amino acids are held together by _____ bonds that are quite stable and can be broken only by _____ or _____ .

23. The curled or spiral arrangement of amino acids in a polypeptide chain is called an _____ . The loops are held in place by _____ between the atoms that are part of the peptide bonds.

24. The zigzag arrangement of the polypeptides in silk is called the _____ . It consists of several parallel polypeptide chains held together by _____ .

25. The geometric arrangements of the amino acids discussed in questions 23 and 24 form the _____ structure of a protein.

26. The long polypeptide chains of globular proteins are tightly folded in an arrangement called the _____ structure of the protein. This folding is a result of interactions between the _____ of the amino acids and may involve one or more of the follow interactions:_____ , _____ , _____ , and _____ .

27. Quaternary structure occurs when the protein has more than one _____ . The interactions involved in this structure may be _____ , _____ , or _____ .

28. Hydrogen bonding is a weak _____ (covalent, noncovalent) linkage, and is easily disrupted by changes in _____ , _____ , _____ or _____ .

29. _____ result when the sulfhydryl groups on the amino acid cysteine undergo oxidation. This is a _____ (covalent, noncovalent) linkage that can be disrupted by _____ , but is stable during changes in _____ , _____ , or _____ .

30. _____ form between the charged groups on the side chains of amino acids such aspartic acid, and are easily disrupted by changes in _____ .

31. _____ form between the R-groups of such amino acids as valine or isoleucine, and occur on the _____ (inside, outside) of the protein molecule.

19.11 Denaturation

A protein can carry out its biological function only when it is in its native state. Denaturation is the disruption of the native state of the protein. Denaturation may or may not be permanent, depending upon the conditions. Changes in pH disrupt hydrogen bonding and salt bridges in proteins, and will cause denaturation. Heat and radiation will increase the motion of a protein molecule, disrupting hydrogen bonding and salt bridges. Small changes in temperature cause reversible denaturation, but high temperatures result in irreversible denaturation and the coagulation of protein. Organic solvents such as alcohol form hydrogen bonds with proteins, altering the existing hydrogen bonding and causing coagulation. Ions of heavy metals such as lead, mercury, or silver may disrupt salt bridges or disulfide bridges, and are toxic to organisms. Alkaloid reagents such as

tannic acid affect salt bridges and hydrogen bonding.   Reducing
agents will break disulfide bridges in a protein.

Important Term

denaturation

Self-Test

For questions 32 to 39, answer true (T) or false (F).

32. When a protein is denatured, the primary structure is
    disrupted.

33. Small changes in pH and in temperature will cause reversible
    denaturation of proteins.

34. Changes in pH affect disulfide bridges in a protein.

35. Strong heating of a protein will cause it to coagulate.

36. Heavy metal ions do their damage to the salt bridges in a
    protein.

37. Eggs are a good antidote for mercury poisoning.

38. Sunburn is a result of denaturation of proteins in the skin.

39. Reducing agents disrupt salt bridges.

Answers to Self-Test Questions in Chapter 19

1. b **2.** c **3.** d **4.** b **5.** c **6.** d **7.** c **8.** a **9.** b **10.** d

11. (a)
$$H_2N-\underset{\underset{\displaystyle CH_3}{|}}{\overset{\displaystyle COOH}{\underset{|}{C}}}-H$$
with $C-CH_3$ and CH_3

(b)
$$H_2N-\underset{\underset{\displaystyle CH_2OH}{|}}{\overset{\displaystyle COOH}{\underset{|}{C}}}-H$$

12. can act as an acid or a base **13.** zwitterions **14.** essential amino acids, adequate, lysine, threonine **15.** isoelectric point, negative, positive **16.** equal to

17. (a) $H_3N^+-\underset{\underset{\displaystyle CH_3}{|}}{CH}-COO^-$

(b) $H_3N^+-\underset{\underset{\displaystyle CH_3}{|}}{CH}-COOH \xleftarrow{+H^+} H_3N^+-\underset{\underset{\displaystyle CH_3}{|}}{CH}-COO^- \xrightarrow{+OH^-} H_2N-\underset{\underset{\displaystyle CH_3}{|}}{CH}-COO^- + H_2O$

18. (a) no charge (b) won't migrate **19.** (a) negative (b) positive **20.** (a) positive (b) negative

21. (a)
$$H_2N-\underset{\underset{\displaystyle CH_3}{\underset{|}{CHCH_3}}}{CH}-\overset{\displaystyle O}{\overset{\|}{C}}-NH-\underset{\underset{\displaystyle CH_3}{|}}{CH}-\overset{\displaystyle O}{\overset{\|}{C}}-NH-\underset{\underset{\displaystyle CH_2OH}{|}}{CH}-\overset{\displaystyle O}{\overset{\|}{C}}-OH$$

(b) Val-Ala-Ser

22. primary, peptide, enzymes, acid or base hydrolysis **23.** alpha helix, hydrogen bonds **24.** beta configuration, hydrogen bonds **25.** secondary **26.** tertiary, R-groups, hydrogen bonding, disulfide bridges, salt bridges, hydrophobic interactions **27.** polypeptide chain, hydrogen bonding, salt bridges, hydrophobic interactions **28.** noncovalent, pH, temperature, solvent, salt concentrations **29.** disulfide bridges, covalent, reducing agents, pH, temperature, salt concentrations **30.** salt bridges, pH **31.** hydrophobic interactions, inside **32.** F **33.** T **34.** F **35.** T **36.** T **37.** T **38.** T **39.** F

Chapter 20   ENZYMES, VITAMINS, AND HORMONES

A major function of globular proteins is to catalyze the reactions occurring in living organisms. Enzymes make up the largest and most highly specialized class of proteins. In this chapter we define the terms used to describe enzymes and their functions. We then discuss the method of enzyme action and the factors that affect this action. Many vitamins are coenzymes, and are essential to the normal functioning of the cell. In the last section of this chapter, we study the ways in which enzyme activity is regulated and inhibited. Hormones play an important role in the regulation of enzyme activity in the body.

20.1 What is Metabolism?

Metabolism refers to all the enzyme-catalyzed reactions that occur in the body. These reactions include anabolic (or biosynthetic) reactions that synthesize molecules needed for repair and growth, and catabolic reactions that break down large molecules to form smaller products and cellular energy.

Important Terms

      metabolism   catabolic reaction   anabolic reaction

20.2 Enzymes

Enzymes are globular proteins that catalyze the reactions of metabolism. Enzyme molecules vary in size and complexity. Because they can be used again and again, they are present in the cell in low concentrations.

The terms used to describe enzymes are defined at the end of this section. Study this terminology carefully before continuing to read the chapter.

Important Terms

enzyme	cofactor
coenzyme	prosthetic group
proenzyme	substrate
active site	holoenzyme
apoenzyme	zymogen

For questions 1 and 2, fill in the blanks with the correct word or words.

1. _____ is the term used to denote the chemical reactions occurring within the body.

2. These reactions can be broken down into two types: _____ reactions that involve the breakdown of large molecules into smaller molecules and chemical energy, and _____ reactions that involve the synthesis of molecules needed by the cell.

For questions 3 to 11, match the statement with the correct answer or answers from column B. An item in column B may be used more than once.

<table>
<tr><td></td><td></td><td>Column B</td></tr>
<tr><td>3.</td><td>Tightly bound cofactor</td><td>A. enzyme</td></tr>
<tr><td></td><td></td><td>B. coenzyme</td></tr>
<tr><td>4.</td><td>Protein portion of the enzyme</td><td>C. apoenzyme</td></tr>
<tr><td></td><td></td><td>D. holoenzyme</td></tr>
<tr><td>5.</td><td>Apoenzyme + cofactor</td><td>E. prosthetic group</td></tr>
<tr><td></td><td></td><td>F. active site</td></tr>
<tr><td>6.</td><td>Inactive enzyme</td><td>G. substrate</td></tr>
<tr><td></td><td></td><td>H. cofactor</td></tr>
<tr><td>7.</td><td>Enzyme acts on this substance</td><td>I. simple protein</td></tr>
<tr><td></td><td></td><td>J. conjugated protein</td></tr>
<tr><td>8.</td><td>Organic molecule as cofactor</td><td>K. zymogen</td></tr>
<tr><td>9.</td><td>Protein catalyst</td><td></td></tr>
</table>

10. Chemical groups other than protein required for enzyme activity.

11. Region on enzyme where substrate attaches.

20.3 Enzyme Nomenclature and Classification

As with the common names for organic compounds, the early nomenclature for enzymes was not at all systematic. In 1961, a system of enzyme classification was proposed that named enzymes by substrate or type of reaction involved. This led to the establishment of six major divisions of enzymes, listed in Table 20.1 of the text.

Important Terms

hydrolases	oxidoreductases	transferases
lyases	isomerases	ligases

For questions 12 to 17, choose the best answer.

12. This class of enzymes catalyze the interconversion of isomers.
    (a) hydrolases   (c) lyases
    (b) transferases   (d) isomerases

13. This class of enzymes catalyzes oxidation-reduction reactions.
    (a) hydrolases   (c) oxidoreductases
    (b) transferases   (d) ligases

14. These enzymes catalyze hydrolysis reactions.
    (a) hydrolases   (c) lyases
    (b) isomerases   (d) ligases

15. With ATP, these enzymes catalyze the formation of new bonds.
    (a) oxidoreductases   (c) lyases
    (b) isomerases   (d) ligases

16. This class of enzymes catalyzes the transfer of functional groups.
    (a) hydrolases   (c) transferases
    (b) isomerases   (d) ligases

17. These enzymes catalyze the elimination of groups to form double bonds.
    (a) oxidoreductases   (c) lyases
    (b) isomerases   (d) ligases

20.4 Method of Enzyme Action

Enzymes lower the activation energy of a reaction by joining with the substrate and increasing the probability that a reaction will occur. (See the series of steps outlined in Section 20.4 in the text.) Each enzyme catalyzes its reaction at a specific rate. The turnover number of an enzyme is the number of substrate molecules transformed per minute by one molecule of enzyme under optimum conditions.

Important Terms

    enzyme-substrate complex   turnover number

20.5 Specificity

Enzymes differ from inorganic catalysts in their specificity.

They might catalyze only one type of reaction or, in extreme
cases, limit their activity to one specific compound.

Important Term

    specificity

20.6 Lock-and-Key Theory

One way to explain the specificity of enzymes is to picture the
geometry of the active site as specially designed for a specific
substrate. Other molecules won't fit in the active site, just as
the configuration of a lock will fit only one key, and other keys
won't turn the lock. Not only must the geometry of the active
site be correct, but the entire enzyme must be in its native
state for it to function properly.

Important Term

    lock-and-key theory

20.7 Induced-Fit Theory

The lock-and-key model explains the action of some enzymes, but
must be modified to fit the action of other enzymes. Enzyme
molecules are flexible. In some cases the active site may not
fit the substrate, but may be induced by the substrate itself to
take on the conformation of the substrate (just as a glove is
induced by a hand to take on the conformation of the hand).

Zymogens or proenzymes are enzymes that have their active sites
blocked. They become active when the site is unblocked—usually
by the hydrolysis of some of the molecule. Cofactors function
either to provide the active site for the enzyme or to form a
bridge between the enzyme and the substrate.

Important Terms

    induced-fit theory   zymogen
    cofactor   proenzyme

Self-Test _____

For questions 18 to 23, fill in the blank with the correct word
or words.

18. The four steps in an enzyme-catalyzed reaction are _____ ,
     _____ , _____ , and _____ .

19. Enzymes increase the rate of a chemical reaction by **lowering** the _____ .

20. The rate at which an enzyme catalyzes a reaction is described by its _____ .

21. If one molecule of the enzyme glutamic dehydrogenase will catalyze the transformation of 500 substrate molecules per second, then its turnover number is _____ .

22. One major difference between enzymes and inorganic catalysts is the enzyme's _____ . This results from the fact that the enzyme has a specific area, called the _____ , to which the substrate attaches.

23. (a) Which drawing, I or II, illustrates the lock-and-key theory?
    (b) Which illustrates the induced-fit theory?

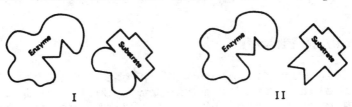

I   II

20.8 Factors Affecting Enzyme Activity

For an enzyme to function properly, its structure must be in the native state. We have seen that changes in the chemical environment, such as changes in pH and temperature, will affect the secondary and tertiary structure of a protein. Each enzyme has an optimum pH at which its activity is greatest, and most body enzymes have maximum activity at normal body temperature. Temperatures and pH's that deviate from these optimum conditions will cause a decrease in the activity of the enzyme. Very high temperatures will permanently denature the enzyme.

Self-Test

For questions 24 to 27, answer true (T) or false (F).

24. The chemical environment of an enzyme is critical to its activity.

25. The digestive enzyme trypsin will have its greatest activity in the stomach.

26. Enzymes are permanently denatured by both low and high temperatures.

27. The activity of the enzyme ribonuclease will increase as the pH changes from 6 to 7.5.

20.9 What Are Vitamins?

Vitamins are organic compounds that the body cannot synthesize, but that are necessary for the normal growth and reproduction of cells. Most vitamins function as coenzymes, and the trace amounts we required must be supplied by the foods we eat. When a particular vitamin is missing in the diet, a specific deficiency disease results. Vitamins are classified as either water soluble or fat soluble.

Important Terms

vitamin	deficiency disease
coenzyme	RDA (Recommended Dietary Allowance)

20.10 Water-Soluble Vitamins

Any water-soluble vitamin absorbed in excess of the daily requirement is eliminated from the body in the urine. Therefore, these vitamins must be present in the diet every day. The water-soluble vitamins include vitamin C and the eight vitamins called the B vitamins. A deficiency in vitamin C causes scurvy; in thiamine, beriberi; in niacin, pellagra; and in vitamin B_{12}, pernicious anemia. Many of the B vitamins are coenzymes in reactions essential to life. Three of these coenzymes serve as hydrogen carriers in the oxidation-reduction reactions that supply cellular energy. Pantothenic acid forms part of the molecule coenzyme A.

Important Terms

water-soluble vitamin	vitamin C	thiamine (B_1)
riboflavin (B_2)	niacin	pyridoxine (B_6)
cyanocobalamin (B_{12})	pantothenic acid	folacin
biotin	scurvy	beriberi
pernicious anemia	pellagra	NAD^+
$NADP^+$	FAD	

20.11 Fat-Soluble Vitamins

The fat-soluble vitamins are vitamins A, D, E, and K. These vitamins are not soluble in water, and are stored in adipose tissue when absorbed into the body in excess of the daily requirements. Although no harmful effects have been shown from high concentrations of water-soluble vitamins, absorption of large amounts of fat-soluble vitamins is very harmful. For example, vitamin D is necessary to prevent the deficiency disease rickets, but too much vitamin D is toxic.

Important Terms

vitamin A	vitamin D	**vitamin E**
vitamin K	rickets	

Self-Test ──

For questions 28 to 41, choose the best answer or answers.

28. Many water soluble vitamins are
    (a) stable
    (b) coenzymes
    (c) heat resistant
    (d) stored in the body

29. Which of the following are fat-soluble vitamins?
    (a) vitamin A
    (b) vitamin B_1
    (c) vitamin D
    (d) vitamin E

30. Deficiency of this vitamin causes scurvy.
    (a) vitamin A
    (b) vitamin B_1
    (c) vitamin C
    (d) vitamin D

31. Low levels of this vitamin cause night blindness.
    (a) vitamin A
    (b) niacin
    (c) thiamine
    (d) vitamin K

32. When there is not enough of this vitamin in the diet the blood fails to coagulate.
    (a) vitamin B_{12}
    (b) folacin
    (c) vitamin E
    (d) vitamin K

33. Which of the following are water-soluble vitamins?
    (a) vitamin B_1
    (b) niacin
    (c) biotin
    (d) vitamin E

34. This forms part of the coenzyme NAD^+.
    (a) vitamin C
    (b) riboflavin
    (c) niacin
    (d) pantothenic acid

35. Lack of this vitamin causes the deficiency disease pernicious anemia.
    (a) vitamin B_1
    (b) vitamin B_2
    (c) vitamin B_6
    (d) vitamin B_{12}

36. This vitamin is important in the regulation of calcium metabolism.
    (a) vitamin D
    (b) vitamin E
    (c) pyridoxine
    (d) thiamine

37. This vitamin forms part of coenzyme A.
   (a) vitamin C   (c) niacin
   (b) riboflavin   (d) pantothenic acid

38. Deficiency of this vitamin causes rickets.
   (a) vitamin A   (c) vitamin D
   (b) vitamin B_1 (d) vitamin E

39. Lack of this vitamin in the diet causes beriberi.
   (a) thiamine   (c) niacin
   (b) riboflavin   (d) pantothenic acid

40. A shortage of this vitamin causes pellagra.
   (a) thiamine   (c) riboflavin
   (b) niacin   (d) vitamin K

41. Any excess of these vitamins is stored in adipose tissue.
   (a) vitamin A   (c) vitamin D
   (b) vitamin C   (d) vitamin E

20.12 Regulatory Enzymes

Cells must have ways to control the activity of enzymes during
the life of the cell. Most events in cellular metabolism require
a series of reactions, called a multienzyme system, each of which
is catalyzed by a specific enzyme. In a multienzyme system there
is a regulatory or allosteric enzyme (usually the enzyme for the
first reaction in the series) that controls the rate of the
entire series. This enzyme is a complex molecule having an
active site and one or more regulatory or allosteric sites. The
activity of this enzyme is controlled (increased or inhibited) by
the presence of substances in the cell called regulatory
molecules.

Quite often these regulatory molecules are the end product of the
reaction series. This type of regulation is called feedback
control. In such a case, high concentrations of the end product
will inhibit the regulatory enzyme. For example, high levels of
ATP in the cell will inhibit the reactions that produce it.
Another way for the cell to control the rate of reactions of the
cell is to control the rate of production of the enzymes.

Important Terms

regulatory enzyme   regulatory site
allosteric site   regulatory molecule
multienzyme system   feedback control
allosteric enzyme

20.13 Hormones

Hormones are chemical substances that coordinate the actions between cells of a multicellular organism. Hormones are produced by a system of glands called the endocrine glands, and influence the activity of particular target organs or cells. The production of hormones may be triggered by the nervous system or by particular chemicals.

Cyclic AMP functions as a secondary messenger or intracellular hormone. Its formation from ATP is triggered by the attachment of a hormone to the membrane of a target cell. Cyclic AMP diffuses throughout the cell, causing the cell to respond to the hormone in a characteristic manner. For example, luteinizing hormone secreted by the pituitary travels through the blood and attaches to target cells in the ovaries, causing the production of cyclic AMP within the cells. The cyclic AMP then stimulates the production and secretion of progesterone by the cells.

Prostaglandins are 20-carbon fatty acids found in almost all cells. They are extremely potent biological agents and, like cyclic AMP, function to carry out messages that cells receive from hormones. Because of their potency, prostaglandins are finding increasing use as therapeutic agents.

Important Terms

    hormone   endocrine gland
    prostaglandin   cyclic AMP

Self-Test _____

For questions 42 to 53, fill in the blank with the correct word or words.

42. A _____ is a sequence of enzyme-catalyzed reactions having a specific metabolic result.

43. The rate at which such a series of reactions occurs is controlled by a _____ , or _____ , enzyme.

44. This enzyme has a complex structure containing an _____ site and one or more _____ sites.

45. A _____ molecule attaches to this second site, causing a change in the enzyme's activity.

46. Many of these systems are controlled by the concentration of the _____ of the reaction. Such control is called _____ .

47. Genes control the rate of chemical reactions within a cell by controlling the rate of enzyme _____ .

48. In multicellular organisms, the activity of enzymes within cells is coordinated by _____ produced by the _____ glands.

49. The production of these chemical messengers is triggered by the _____ system or specific _____ .

50. Hormones control the activity within the cell by _____ or _____ .

51. _____ and _____ are produced in the cell and function as secondary messengers within the cell. Their production is triggered by a _____ attaching to the cell membrane.

52. _____ is produced from ATP when the enzyme adenyl cyclase is activated.

53. _____ are potent chemicals that are used therapeutically to induce labor and relieve asthma.

20.14 Irreversible Inhibition

Irreversible inhibition of an enzyme occurs when a functional group or a cofactor required for the enzyme's activity is destroyed or permanently modified.

Important Term

irreversible inhibition

20.15 Reversible Inhibition

Competitive inhibition occurs when a compound whose structure is similar to that of the substrate competes with the substrate for the active site, thus decreasing the activity of the enzyme. Noncompetitive inhibition involves the inhibitor combining reversibly with a portion of the enzyme (other than the active site) that is essential to the activity of the enzyme.

Antimetabolites are a class of compounds that inhibit enzyme action. Antibiotics are antimetabolites. They work by preventing the growth of microorganisms. Antibiotics function in many ways: inhibiting the formation of the bacterial cell wall, increasing the permeability of the cell membrane, or interfering with protein synthesis within the cell. Chemicals used in cancer chemotherapy are often antimetabolites that compete for the active site on an enzyme in the cancerous cell.

Important Terms

 reversible inhibition   competitive inhibition
 noncompetitive inhibition   antimetabolite
 antibiotic   chemotherapy

Self-Test _____

For questions 54 to 58, fill in the blank with the correct word
or words.

54. When an inhibitor binds tightly to a metal ion cofactor,the
    inhibition is _____ .

55. Inhibition caused by a molecule whose shape closely
    resembles that of the substrate is called _____
    inhibition.

56. Inhibition caused by a compound that binds reversibly with
    the sulfhydryl groups on the enzyme is called _____
    inhibition.

57. An antimetabolite is a compound that _____ .

58. _____ are antimetabolites that are extracted from cultures
    of microorganisms.

For questions 59 to 64, match the statement with the correct
answer or answers from column B.

		Column B
59.	Inhibits cholinesterase	A. iodoacetic acid
		B. lead ion
60.	Antimetabolite used in cancer therapy	C. 5-fluorouracil
		D. nerve gas
61.	Binds irreversibly to sulfhydryl groups	E. cephalosporin
		F. sulfanilamide
62.	Binds reversibly to sulfhydryl groups	G. mercury ion
		H. penicillin
63.	Inhibits the formation of bacterial cell walls.	I. malonic acid
64.	Competes with succinic acid for the active site on succinate dehydrogenase	

Answers to Self-Test Questions in Chapter 20

1. metabolism 2. catabolic, anabolic 3. E 4. C 5. D,J 6. K
7. G 8. B 9. A 10. H 11. F 12. d 13. c 14. a 15. d 16. c
17. c 18. (a) The enzyme-substrate complex forms (b) The
substrate becomes activated (c) The products form on the surface
of the enzyme (d) The products are released from the surface of
the enzyme 19. activation energy 20. turnover number
21. 3×10^4 molecules/min 22. specificity, active site
23. (a) II (b) I 24. T 25. F 26. F 27. T 28. b 29. a,c,d
30. c 31. a 32. d 33. a,b,c 34. c 35. d 36. a 37. d 38. c
39. a 40. b 41. a,c,d 42. multienzyme system 43. regulatory,
allosteric 44. active, regulatory (allosteric) 45. regulatory
46. end product, feedback control 47. synthesis 48. hormones,
endocrine 49. nervous, chemical compounds 50. entering the cell
and combining with a protein, attaching to the cell membrane
51. cyclic AMP, prostaglandins, hormone 52. cyclic AMP
53. prostaglandins 54. irreversible 55. competitive
56. noncompetitive 57. inhibits enzyme action 58. antibiotic
59. D 60. C 61. G 62. A 63. H 64. I

Chapter 21   PATHWAYS OF
                METABOLISM

In this chapter we study the way in which the body digests
carbohydrates, lipids, and proteins. We look at the cellular
energy requirements, and how the cell uses molecules produced in
the digestion of these food groups for synthesis of other
compounds and for the production of energy.

21.1 Cellular Energy Requirements

The compounds used by the body to meet the immediate energy needs
of cells are "high-energy" phosphates such as ATP. ATP contains
two oxygen-to-phosphorus bonds that give off large amounts of
energy when hydrolyzed. The hydrolysis of ATP supplies the
energy needed for the anabolic reactions in the cell.

Important Terms

        adenosine triphosphate, ATP   "high-energy" phosphate bond
        adenosine diphosphate, ADP   adenosine monophosphate, AMP

Self-Test

For questions 1 to 4, fill in the blank with the correct word or
words.

  1. Compounds that contain _____ bonds supply the energy
      needed for endothermic reactions in the cell.

  2. One energy-rich compound is _____ .

  3. In this compound, energy is released when a _____ -to-
      _____ bond is hydrolyzed.

  4. Hydrolysis of a high-energy phosphate bond releases _____
      kcal of energy. An ordinary phosphate bond releases about
      _____ kcal.

21.2 Digestion and Absorption

Digestion is the process by which the body breaks down food so
that it can be absorbed and used. Digestion of carbohydrates
begins in the mouth, where enzymes in saliva begin the breakdown

of starch. The remainder of carbohydrate digestion occurs in the small intestine, where enzymes in the pancreatic and intestinal juices break polysaccharides and disaccharides into the monosaccharides glucose, fructose, and galactose. These monosaccharides are absorbed through the intestinal wall into the blood and carried to the cells, where they enter various metabolic pathways.

The digestion of lipids begins in the stomach, where stomach acid begins the hydrolysis of fat. The majority of lipid digestion, however, takes place in the small intestine, where bile salts emulsify the fats so they can be broken down by enzymes in the pancreatic and intestinal juices. Bile is a fluid manufactured by the liver and stored in the gall bladder before being released into the small intestine. It contains cholesterol, bile salts, and bile pigments. Fats are broken down to glycerol, fatty acids, and mono- and diacylglycerols, which cross the intestinal barrier and are carried to the blood via the lymph system. They are carried in the blood as micro-droplets joined to proteins called chylomicrons.

Digestion of proteins begins in the stomach, where enzymes activated by stomach acid break proteins down into polypeptides. The polypeptides are then hydrolyzed in the small intestine by enzymes in the pancreatic and intestinal juices to amino acids and small peptides. These products are absorbed into the bloodstream.

Important Terms

      digestion   bile   proenzyme

Self-Test _____

For questions 5 to 15, answer true (T) or false (F).

5. Many carbohydrates found in our food must be broken down into smaller molecules before they can be used by the cells in our body.

6. Digestion of carbohydrates begins in the stomach.

7. Lactose and maltose formed in the small intestine can be absorbed into the bloodstream.

8. Glucose formed in digestion may be stored in the liver as glycogen until needed.

9. Digestion of fats takes place in the stomach.

10. Gall stones are crystallized bile.

11. Bile pigments are formed from the breakdown of hemoglobin.

12. Bile is produced and stored in the gall bladder.

13. Jaundice is caused by too much cholesterol in the blood.

14. Protein digestion begins in the mouth, where enzymes in saliva begin protein hydrolysis.

15. Protein-digesting enzymes are released in the stomach as zymogens or proenzymes.

21.3 Glycogenesis and Glycogenolysis

Glycogenesis is the process by which glucose is polymerized to form glycogen in the liver when excess glucose is present in the blood. When the blood sugar level is low, glycogen is broken down in the liver by a series of reactions called glycogenolysis.

Insulin is produced by the beta cells in the pancreas when the concentration of glucose in the blood is high. It accelerates the absorption of glucose from the blood by muscle and fatty tissue, increases the production of glycogen, and decreases glycogenolysis in the liver.

The normal functioning of the brain requires a fairly constant level of glucose in the blood. Hypoglycemia is a condition caused by abnormally low concentrations of blood glucose. It can result in dizziness, lethargy, fainting, and (in severe cases) shock and coma. Hyperglycemia results from abnormally high concentrations of glucose in the blood, and occurs in conditions such as stress, kidney failure, and diabetes mellitus. When the glucose level exceeds that tolerated by the kidneys, glucose will enter the urine.

Important Terms

glycogenesis	glycogenolysis
hypoglycemia	hyperglycemia
insulin shock	insulin
glucagon	hyperinsulinism
renal threshold	glycosuria

21.4 Oxidation of Carbohydrates

The glucose in our blood enters the cells and is either used in biosynthetic reactions or is oxidized to carbon dioxide, water, and energy in the form of ATP in a process called cellular respiration. When we compare the energy produced by the oxidation of glucose in the laboratory with the energy produced by the oxidation of glucose in our cells, we find that our cells are able to trap about 44% of this energy; the rest of the energy is lost as heat to maintain body temperature.

The oxidation of glucose can be divided into two stages. The first stage, called glycolysis, is anaerobic (requires no oxygen.) It occurs in the cytoplasm of the cell, and produces two molecules of lactic acid and two molecules of ATP.

The second stage in the oxidation of glucose is aerobic (requires oxygen.) It occurs in the mitochondria of the cell, serves as the final stage of oxidation of other fuel molecules, and produces carbon dioxide, water, and ATP. In this stage, there are two series of reactions: the citric acid cycle and the respiratory chain.

Important Terms

cellular respiration   glycolysis
citric acid cycle   aerobic
respiratory chain   anaerobic
mitochondria

21.5 Citric Acid Cycle

The citric acid cycle is a cyclic series of reactions in which acetic acid (in the form of acetyl-CoA produced by the oxidation of pyruvic acid) is oxidized to two molecules of carbon dioxide, four pairs of hydrogen atoms, and one GTP. The hydrogen atoms are carried to the respiratory chain by NAD^+ and FAD. The reactions of the citric acid cycle are shown in Figure 21.8 in the text, and you should be able to follow its cyclic nature. The first reaction is the coupling of acetyl-CoA with a molecule of oxaloacetate, and the last reaction produces a molecule of oxaloacetate to be used in another turn of the cycle.

Important Terms

coenzyme A   acetyl-CoA

21.6 Respiratory Chain and ATP Formation

The four pairs of hydrogens produced in the oxidation of acetyl-CoA in the citric acid cycle enter the respiratory chain, where first these atoms, and then their electrons, are passed between a series of compounds. Each transfer involves an oxidation-reduction reaction in which energy is released. The exact mechanisms of the production of ATP in this process are not understood, but the points at which ATP is formed along the chain have been determined. Oxygen must be present for the respiratory chain to occur, and the products of this series of reactions are ATP and water. The citric acid cycle and the respiratory chain occur together, and anything that disrupts either series of reactions will disrupt the entire process of aerobic oxidation.

21.7 Lactic Acid Cycle

During strenuous exercise the body cannot supply the muscle cells with enough oxygen, so the cells must rely on glycolysis for the production of energy. Glycolysis can occur until the lactic acid that is produced hampers muscle performance, causing muscle fatigue and exhaustion. This lactic acid also enters the blood, causing acidosis and its complications. In order for the body to recover, the lactic acid must be oxidized by the muscle cells or liver, or be converted to glycogen by the liver. Heavy breathing after exercise supplies the oxygen necessary for these reactions.

Important Terms

lactic acid cycle   lactic acid   acidosis

21.8 Fermentation

Yeast obtain all the energy they need by glycolysis, but the end product of this glycolysis is alcohol rather than lactic acid. The name given to this form of glycolysis is fermentation.

Important Term

fermentation

Self-Test _____

For questions 16 to 23, fill in the blank with the correct word or words.

16. The process of converting glucose to glycogen is called
    _____ . The breakdown of glycogen to glucose is called
    _____ . The rates of these two reactions series are
    controlled by the hormones _____ and _____ .

17. _____ controls the rate at which glucose in the blood
    enters the cells of the muscles and fatty tissue. It is
    synthesized by the _____ in the pancreas when the blood

sugar level is _____ (high, low) and acts to _____ (lower, raise) the concentration of glucose in the blood.

18. The normal fasting level of glucose in the blood of an adult is _____ per 100 mL of blood.

19. A blood sugar concentration below normal is called _____. Its symptoms in mild cases are _____ , and in severe cases _____ .

20. Overinjection of insulin results in a condition called _____, and can cause convulsions and coma referred to as _____ .

21. A meal high in _____ can cause overproduction of insulin and lowering of the blood sugar level.

22. A high concentration of glucose in the blood is called _____ and can be caused by _____ .

23. When the level of glucose is above the _____ , glucose will be excreted in the urine.

For questions 24 to 28, answer true (T) or false (F).

24. Glycogenesis and glycolysis involve exactly opposite reaction pathways.

25. Oxidation of glucose in cells is exothermic, and all of the energy is trapped in ATP molecules.

26. The reactions of the citric acid cycle and the respiratory chain cannot occur separately.

27. When sufficient oxygen is present, pyruvic acid is oxidized to acetic acid rather than lactic acid.

28. The majority of ATP formed in the oxidation of glucose is produced in the respiratory chain.

For questions 29 to 45, choose the best answer or answers.

29. Reactions that require oxygen are called
    (a) anaerobic   (c) anabolic
    (b) aerobic   (d) catabolic

30. Another name for the Krebs cycle is the
    (a) citric acid cycle   (c) lactic acid cycle
    (b) respiratory chain   (d) fermentation

31. A series of reactions in which the end products are water and ATP.
    (a) citric acid cycle   (c) lactic acid cycle
    (b) respiratory chain   (d) fermentation

32. A substance produced in muscles during strenuous exercise.
    (a) glucose   (c) lactic acid
    (b) glycogen   (d) pyruvic acid

33. The end products of this reaction are alcohol and carbon dioxide.
    (a) glycogenesis   (c) lactic acid cycle
    (b) glycolysis   (d) fermentation

34. The enzymes of the citric acid cycle are found here.
    (a) nucleus   (c) chloroplast
    (b) mitochondria   (d) cytoplasm

35. The process by which glycogen is synthesized from glucose.
    (a) glycogenesis   (c) glycolysis
    (b) glycogenolysis   (d) fermentation

36. Reactions that can take place in the absence of oxygen are called
    (a) anaerobic   (c) anabolic
    (b) aerobic   (d) catabolic

37. The series of reactions by which acetyl-CoA is oxidized to carbon dioxide.
    (a) citric acid cycle   (c) lactic acid cycle
    (b) respiratory chain   (d) fermentation

38. A series of redox reactions that release energy that is then trapped in molecules of ATP.
    (a) citric acid cycle   (c) lactic acid cycle
    (b) respiratory chain   (d) glycolysis

39. This compound causes muscle fatigue.
    (a) pyruvic acid   (c) acetyl-CoA
    (b) citric acid   (d) lactic acid

40. The reactions of glycolysis occur here.
    (a) nucleus   (c) cytoplasm
    (b) mitochondria   (d) chloroplast

41. The breakdown of glycogen to glucose occurs in
    (a) glycogenesis   (c) glycolysis
    (b) glycogenolysis   (d) fermentation

42. The series of reactions that involves the oxidation of glucose to lactic acid.
    (a) citric acid cycle   (c) lactic acid cycle
    (b) respiratory chain   (d) glycolysis

43. The molecule that carries acetic acid to the citric acid cycle.
    (a) pyruvic acid   (c) NAD$^+$
    (b) coenzyme A   (d) FAD

44. Hydrogen carriers in the electron transport chain.
    (a) pyruvic acid   (c) NAD$^+$
    (b) coenzyme A   (d) FAD

45. When glycolysis occurs in yeast it is called
    (a) glycogenesis   (c) oxidation
    (b) glycogenolysis   (d) fermentation

21.9 Lipogenesis

Adipose tissue contains triacylglycerols formed when excess
lipids and carbohydrates are consumed. The synthesis of these
lipids is called lipogenesis. In addition to storing energy,
adipose tissue functions as a support for inner organs and a heat
insulator. Adipose tissue is not stable. It is constantly being
synthesized and broken down as stored fatty acids are exchanged
for food fatty acids.

Important Terms

    lipogenesis   adipose tissue

21.10   Oxidation of Fatty Acids

Fats yield more energy per gram than carbohydrates. Oxidation of
triacylglycerols begins when they are hydrolyzed to glycerol and
fatty acids. The glycerol enters the glycolysis pathway. The
fatty acids are oxidized in a sequence of reactions called the
fatty acid cycle or beta oxidation. These reactions (shown in
Figure 21.12 in the text) are cyclic, and begin with the
oxidation of the beta carbon. With each turn of the cycle, one
acetyl-CoA and one pair of hydrogens are produced.
The hydrogens enter the respiratory chain, and the acetyl-CoA's
may enter the citric acid cycle or be used in the synthesis of
other compounds.

Important Terms

    beta oxidation   fatty acid cycle

21.11   Metabolic Role of Acetyl-CoA

Acetyl-CoA plays a central role in cellular metabolism.   In addition to being the starting compound for the citric acid cycle, it is the starting material for the synthesis of compounds such as fatty acids, amino acids, and cholesterol. The body rids itself of excess acetyl-CoA by producing ketone bodies.

21.12 Ketone Bodies

When an abnormal increase in the level of fat metabolism occurs, excess acetyl-CoA is produced. This acetyl-CoA is converted in the liver to three compounds called ketone bodies. When the level of these compounds is high, ketosis results. Ketone bodies will be found in the urine and will cause acidosis in the blood. In extreme cases this can lead to coma and death.

Important Terms

ketone bodies   ketosis   acidosis

Self-Test _____

For questions 46 to 51, answer true (T) or false (F).

46. Adipose tissue contains stored triacylglycerols.

47. The composition of adipose tissue is determined by the type of diet.

48. The oxidation of a gram of carbohydrate will yield about twice as much energy as a gram of fat.

49. In the beta oxidation of fatty acids, four-carbon units are removed with each turn of the cycle.

50. The importance of acetyl-CoA in cellular metabolism is that it is the compound that enters the citric acid cycle.

51. The formation of ketone bodies occurs when the rate of fat metabolism is reduced.

For questions 52 to 56, fill in the blank with the correct word or words.

52. The oxidation of fats begins with the _____ of a triacylglycerol. The glycerol formed is oxidized in the _____ pathway, and the fatty acids are oxidized by the _____ .

53. In one turn of the fatty acid cycle _____ , _____ . and _____ are produced.

54. The complete oxidation of behenic acid ($CH_3(CH_2)_{20}COOH$) will produce _____ ATPs.

55. Ketone bodies are found in the urine in a condition called _____ . In this condition, the pH of the blood will _____ (increase, decrease). This condition is caused by an increase in the level of _____ metabolism.

56. An untreated diabetic can lapse into a coma caused by _____. This severe condition results because the diabetic must oxidize _____ to provide energy for the cells.

21.13 Metabolic Uses of Amino Acids

The amino acids that enter the blood are used in the synthesis of new proteins and nonprotein, nitrogen-containing compounds. Amino acids that are absorbed in excess of the synthetic needs of the body are catabolized. The amino group is converted into urea and the remainder of the molecule enters the citric acid cycle or is used to synthesize other amino acids, glucose, and glycogen.

There is no storage form for amino acids in the body. Rather, the body maintains a pool of amino acids that constantly changes as protein is broken down and resynthesized.

21.14 Amino Acid Synthesis

Amino groups can be transferred from amino acids to form new amino acids in a process called transamination. In this series of reactions, an amino group is transferred from an amino acid to an α-keto acid, producing a new amino acid and a new α-keto acid.

Amino acids are used in the synthesis of many important compounds such as hormones, skin pigments, nerve transmitters, coenzymes, lipids, and bile salts.

Important Terms

transamination   α-keto acid

21.15 Amino Acid Catabolism: Oxidative Deamination

Amino acids not used by the cell in synthetic reactions are catabolized to ammonia, carbon dioxide, water, and energy. The first step in this process is oxidative deamination, in which the amino group is removed from an amino acid to produce an α-keto acid and ammonia. The α-keto acid can enter the citric acid cycle, be used to produce other amino acids, or be used in the production of carbohydrates or fats. The ammonia produced enters the urea cycle.

Important Term

oxidative deamination

21.16 Urea Cycle

Ammonia is quite toxic to cells and cannot be allowed to build up within the body. Cells of the liver convert ammonia to the less toxic urea, which can be concentrated by the kidneys and excreted from the body. Figure 21.16 in the text shows the cyclic series of reactions of the urea cycle. Any disruption in the excretion of urea can be fatal.

Important Terms

urea   urea cycle   uremia

21.17 Pathways of Metabolism

The metabolism of carbohydrates, lipids, and proteins is intricately interrelated. A disruption in any of these pathways will have profound effects on all aspects of metabolism within the human body.

Self-Test _____

For questions 57 to 63, fill in the blank with the correct word or words.

57. Nitrogen used in synthesis reactions in the cells is obtained from _____ that we eat.

58. The largest amounts of protein in the diet are needed during periods of _____ or _____ .

59. The body synthesizes new amino acids by a process called _____ .

60. In amino acid catabolism, amino acids are broken down first by a process called _____ . This reaction produces an _____ acid that can enter the _____ cycle or be used to synthesize_____, _____, or _____ .

61. The other product of the reaction in question 60 is _____. It enters the _____ cycle, which takes place in cells of the _____ .

62. A blockage of the kidneys can cause a build-up of urea and other wastes in the blood, resulting in a condition called _____. These wastes can be removed from the patient's blood by _____ .

63. When dieting it is important to maintain a certain minimum level of _____ in the diet to prevent the breakdown of tissue proteins and the formation of ketone bodies.

64. Write equations for the following reactions:
    (a) The transamination reaction between valine and α-ketoglutaric acid
    (b) The deamination of valine
    (c) The overall reaction of the urea cycle.

Answers to Self-Test Questions in Chapter 21

1. high energy phosphate   2. ATP (UTP,GTP)   3. oxygen, phosphorus
4. 7, 2-5   5. T   6. F   7. F   8. T   9. T   10. T   11. T   12. F
13. F   14. F   15. T   16. glycogenesis, glycogenolysis, insulin,
glucagon   17. insulin, beta cells, high, lower   18. 60-100 mg
19. hypoglycemia, irritability and grogginess, convulsions and
coma   20. hyperinsulinism, insulin shock   21. refined
carbohydrate   22. hyperglycemia, diabetes mellitus, stress,
kidney failure, certain drugs   23. renal threshold   24. F   25. F
26. T   27. T   28. T   29. b   30. a   31. b   32. c   33. d   34. b
35. a   36. a   37. a   38. b   39. d   40. c   41. b   42. d   43. b
44. c,d   45. d   46. T   47. T   48. F   49. F   50. F   51. F
52. hydrolysis, glycolysis, beta oxidation (fatty acid cycle)
53. acetyl-CoA, NADH + H$^+$,   FADH$_2$ 54. 181   55. ketosis,
decrease, fat   56. ketosis (acidosis), fats   57. protein
58. growth and cellular repair   59. transamination   60. oxidative
deamination, α-keto, citric acid, amino acids, carbohydrates,
fats   61. ammonia, urea, liver   62. uremia, hemodialysis
64. carbohydrate

65.   (a)

$$\underset{\substack{\big| \\ CH-CH_3 \\ \big| \\ CH_3}}{\overset{\substack{COOH \\ \big|}}{H-C-NH_2}} + \underset{\substack{\big| \\ CH_2 \\ \big| \\ CH_2 \\ \big| \\ COOH}}{\overset{\substack{COOH \\ \big|}}{C=O}} \longrightarrow \underset{\substack{\big| \\ CH-CH_3 \\ \big| \\ CH_3}}{\overset{\substack{COOH \\ \big|}}{C=O}} + \underset{\substack{\big| \\ CH_2 \\ \big| \\ CH_2 \\ \big| \\ COOH}}{\overset{\substack{COOH \\ \big|}}{H-C-NH_2}}$$

(b)

$$\underset{\substack{\big| \\ CH-CH_3 \\ \big| \\ CH_3}}{\overset{\substack{COOH \\ \big|}}{H-C-NH_2}} \xrightarrow[\underset{NAD^+ \quad\quad NADH + H^+}{}]{H_2O} \underset{\substack{\big| \\ CH-CH_3 \\ \big| \\ CH_3}}{\overset{\substack{COOH \\ \big|}}{C=O}} + NH_3$$

(c) $2NH_3 + CO_2 \longrightarrow H_2N-\overset{\overset{\textstyle O}{\|}}{C}-NH_2 + H_2O$

Chapter 22   NUCLEIC ACIDS

Whether a cell will develop into a one-celled amoeba or a multibillion-celled human being is determined by the sequence of nitrogen-containing bases in a molecule called deoxyribonucleic acid, or DNA. This molecule belongs to a class of compounds called nucleic acids. In this chapter, we discuss the structure of nucleic acids, their method of replication, and the steps in protein synthesis. We see how errors in the DNA molecule can cause disease and death, and we discuss the molecular basis of the disease PKU.

22.1 Molecular Basis of Heredity

The specific characteristics of an organism are determined by its genes. 'Genes are segments of the DNA molecule that contain the information necessary to make one polypeptide chain. Nucleic acids, then, form the molecular basis of heredity because they control the nature of the proteins synthesized by the cell.

22.2 Nucleotides

DNA (deoxyribonucleic acid) and RNA (ribonucleic acid) are both nucleic acids, which are polymers of nucleotides. When hydrolyzed, nucleotides yield three components: a nitrogen-containing base, a five-carbon sugar, and phosphoric acid. The nitrogen bases are heterocyclic ring compounds that can be classified as pyrimidines or purines. The pyrimidines are uracil, thymine, and cytosine. The purines are adenine and guanine. Uracil is found only in the nucleotides of RNA, and thymine in the nucleotides of DNA. DNA contains the sugar deoxyribose, and RNA the sugar ribose.

In addition to forming the building blocks of nucleic acids, nucleotides serve as coenzymes and as high-energy phosphate molecules. The arrangement of nucleotides in a DNA molecule determines the genetic information carried by that molecule.

Important Terms

nucleic acid	nucleotide	pyrimidine
purine	uracil (U)	thymine (T)
cytosine (C)	guanine (G)	adenine (A)
ribose	deoxyribose	DNA
RNA		

22.3 The Structure of DNA

In 1953 Watson and Crick described a molecule of DNA as two helical polynucleotide chains coiled around a common axis, with the nitrogen-containing bases toward the inside of the helix and the deoxyribose and the phosphate groups on the outside of the helix.

The backbone of the helix is a chain of deoxyribose sugars and phosphate groups connected by ester bonds. Look closely at Figure 22.3 in the text. The base on each nucleotide extends toward the inside of the double helix and is joined by hydrogen bonding to a base on the other strand. The pairing is very specific: adenine pairs only with thymine, and guanine only with cytosine. This means that the two strands of DNA are not identical, but rather are complementary.

Important Term

> double helix

22.4 Replication of DNA

For genetic information to be passed on when a cell divides, the DNA molecules in the cell's nucleus must make identical copies of themselves—they must replicate. In this process, the two strands of the helix unwind and each serves as a template for the formation of a new strand. Each of the new double helixes produced will have one original DNA strand and one newly-made strand.

The genetic information carried by the DNA molecule is contained in the sequence of nitrogen bases in the molecule. A change in this order of bases will result in a mutation in the genes in the cell. The enzymes that replicate DNA have elaborate control mechanisms to check for and repair any errors in the newly synthesized DNA chains.

Important Terms

> replication   mutation

Self-Test _____

For questions 1 to 13, fill in the blank with the correct word or words.

1. Genes are segments of molecules of _____, and carry the information required to produce one _____ chain.

2. _____ are polymers of nucleotides.

3. A nucleotide is composed of three subunits: _____, _____, and _____ .

4. Thymine and cytosine are in the class of bases called _____, and adenine and guanine are in the class of bases called _____ .

5. RNA contains the sugar _____, and DNA contains the sugar _____ .

6. A DNA molecule contains _____ polynucleotide chains, which form a _____ .

7. The bases in a DNA molecule are directed toward the _____ of the molecule. The _____ and _____ make up the backbone of the helix.

8. The strands of the DNA molecule are not identical, but are _____ .

9. The bases of the two strands of DNA are joined by _____ ; the base cytosine will join only with _____, and adenine only with _____ .

10. When a DNA molecule produces another molecule identical to itself, it is undergoing the process called _____ .

11. In this process, the two strands of the DNA _____, and each strand serves as a _____ for the production of a new strand. Each daughter helix will contain one _____ strand and one _____ strand.

12. The genetic information is contained in the _____ on the molecule of DNA.

13. Anything that causes a change in this sequence of bases causes a _____ in the gene.

14. (a) Draw the structure of the nucleotide containing the base adenine and the sugar deoxyribose.
    (b) Draw the structure of the nucleotide containing the base uracil and the sugar ribose.
    (c) Would the nucleotide in (a) or (b) be found in RNA?

15. The following is a sequence of bases found on one strand of a DNA molecule. What would be the sequence of bases on the other DNA strand of the helix?

A-G-G-T-A-C-G-A-C

16. List three functions of nucleotides in the cell.

22.5 Messenger RNA

RNA consists of a single strand of nucleic acid containing the sugar ribose and the bases adenine, cytosine, guanine, and uracil. mRNA is synthesized on the DNA strand in the nucleus of the cell in a process called transcription. It then migrates to the ribosomes in the cytoplasm where mRNA serves as a template for protein synthesis.

Important Term

messenger RNA (mRNA)   transcription

22.6 Ribosomal RNA

rRNA is synthesized on a DNA template in the nucleoli of the nucleus. rRNA joins with protein to form the two subunits of the ribosomes, the sites of protein synthesis.

Important Terms

ribosomal RNA (rRNA)   ribosomes

22.7 Transfer RNA

tRNA, synthesized in the nucleus on DNA, is the smallest and most mobile of the RNAs. Each tRNA is designed to attached to a specific amino acid, thereby becoming "charged". This tRNA-amino acid complex migrates to the ribosomes where the amino acid is added to the end of the growing polypeptide chain.

Important Term

transfer RNA (tRNA)

Self-Test _____

For questions 17 to 23, choose the best answer or answers.

17. This molecule joins with proteins to form the ribosomes.
    (a) DNA   (c) rRNA
    (b) mRNA   (d) tRNA

18. This molecule serves as a template for the sequence of amino acids in protein synthesis.
    (a) DNA   (c) rRNA
    (b) mRNA   (d) tRNA

19. This molecule carries amino acids to the ribosomes.
    (a) DNA   (c) rRNA
    (b) mRNA   (d) tRNA

20. If an amino acid attaches to the terminal adenine
    nucleotide, the result is a
    (a) replicated DNA   (c) charged tRNA
    (b) translated RNA   (d) mutated DNA

21. The site of protein synthesis in the cell is the
    (a) ribosome   (c) nucleus
    (b) nucleolus   (d) cytoplasm

22. The smallest RNA molecule is
    (a) mRNA   (c) tRNA
    (b) rRNA   (d) hnRNA

23. The first step in transmission of genetic information from a
    gene to a polypeptide chain is
    (a) replication of DNA   (c) charging of tRNA
    (b) translation of mRNA   (d) transcription of DNA to mRNA

24. What would be the sequence of bases on a mRNA molecule
    synthesized on the DNA template shown below?

                    A-A-T-G-A-G-C-C-T

22.8   The Genetic Code

The genetic code is the term used to describe the specific
sequence of bases found in a gene. A gene is a segment of a DNA
molecule that will determine the sequence of amino acids in one
polypeptide chain. A sequence of three bases on mRNA, called a
codon, is necessary to code for each amino acid. Table 22.1 in
the text shows these codons. Three codons do not code for any
amino acid, but instead are terminal codons signaling the end of
the polypeptide chain.

An anticodon is a three-base sequence on transfer RNA that is
complementary to the codon on the messenger RNA. Each tRNA
molecule contains a region that carries the anticodon.

Important Terms

        codon   terminal codon   anticodon

236

25. The following are segments of three mRNA molecules. Mark off the codons on each mRNA and determine the sequence of amino acids for which it codes.

    (a) U-U-C-U-C-A-A-C-U-G-A-U
    (b) G-U-U-G-C-A-A-G-A-A-A-A
    (c) A-U-C-C-C-A-C-A-C-C-U-G

22.9 Introns and Exons

Until recently, all of the scientific information on the transcription and translation of genetic information had come from studying one-celled bacteria having no nucleus. In these cells, mRNA is synthesized on a single strand of DNA and then combines with the ribosomes to synthesize proteins.

However, it has been discovered that in eukaryotic cells (cells with nuclei) the RNA synthesized on the DNA strand is much longer than the mRNA that exits from the nucleus. This RNA, called heterogeneous nuclear RNA (hnRNA), contains sequences of bases called exons that code for amino acids, and intervening sequences of bases called introns. After the hnRNA is synthesized on the DNA, special enzymes in the nucleus cut the RNA and splice out all of the introns, producing the mRNA that migrates from the nucleus to the ribosomes in the cytoplasm.

Important Terms

　　heterogeneous nuclear RNA (hnRNA)　　　exon　　　intron

22.10 The Steps in Protein Synthesis

The first step in protein synthesis is the synthesis of a strand of mRNA—the transcription of DNA to mRNA. Next occurs the translation of the mRNA to protein. This sequence of reactions involves the attachment of amino acids to tRNA, the starting of a polypeptide chain, the addition of amino acids to the polypeptide chain, and the termination of the polypeptide chain. Many different ribosomes can be synthesizing protein on the same strand of mRNA, and such groups of ribosomes are called polysomes.

Important Terms

　　transcription of DNA　　　translation of mRNA　　　polysome

For questions 26 to 30, fill in the blank with the correct word or words.

26. The synthesis of mRNA on DNA is called _____, and the synthesis of a protein on the mRNA is called _____ .

27. In eukaryotic cells, _____ RNA, which is much longer than mRNA, is synthesized on the DNA. Special enzymes in the nucleus splice out the _____, which are sequences of bases that are between the _____, the sequences of bases that code for amino acids.

28. Before the synthesis of a protein can begin, _____ must be synthesized in the nucleus, and tRNA must become _____ .

29. To begin the synthesis of a protein, the _____ subunit of a ribosome attaches to mRNA, and a tRNA with an specific initiating amino acid. This complex then joins with the _____ subunit to form the ribosome.

30. The addition of amino acids to the polypeptide chain stops when a _____ is reached on the mRNA.

22.11 Recombinant DNA Technology

Recombinant DNA is a laboratory technique that is allowing scientists to study the mysteries of the cell. It is also allowing for the commercial production of human hormones, antibiotics, and vaccines. This technique splices specific segments of DNA onto the DNA of E.coli , which will then produce the desired protein.

Important Terms

recombinant DNA   molecular cloning   genetic engineering

22.12 Mutations

A mutation is a change in the sequence of bases on the DNA molecule. Such changes can occur spontaneously, or can be caused by viruses, radiation, or chemicals. The abnormal proteins that result from mutations can cause disorders, disease, or death to the organism. Over 2000 diseases, such as sickle cell anemia, hemophilia, and PKU, result from mutations in genes and the defective enzymes produced by these mutations.

Important Term

   mutation

22.13   Phenylketonuria

Phenylketonuria (PKU) is a disease resulting from a mutation in
the gene that codes for the liver enzyme phenylalanine
hydroxylase. Phenylalanine hydroxylase normally catalyzes the
conversion of the amino acid phenylalanine to the amino acid
tyrosine. The defective enzyme that is produced in individuals
inheriting PKU genes from both parents will not catalyze this
reaction. As a result, phenylalanine and PKU metabolites,
substances the body produces from phenylalanine in an attempt to
rid itself of the excess, build up in the body. This produces an
abnormal chemical environment for body cells, causing
irreversible damage when PKU is untreated.

Important Terms

   phenylketonuria, PKU   PKU metabolites

Self-Test _____

For questions 31 to 39, match the statement with the correct
answer or answers in column B. An item in column B may be used
more than once.

<u>Column B</u>

31. This enzyme is defective in PKU   A. transcription
                                        B. translation
32. PKU metabolite   C. mutation
                                        D. phenylacetic acid
33. A change in the base sequence   E. phenylalanine
     of DNA   F. recombinant DNA
                                        G. phenylalanine
34. Disease producing an   hydroxylase
     accumulation of phenylalanine   H. PKU
                                        I. tyrosine
35. DNA ⟶ mRNA

36. mRNA ⟶ protein

37. Phenylalanine ⟶ tyrosine

                   ultraviolet
38. T-C-A-G-C-A ⟶ T-C-T-G-C-A
                      light

39. Uses <u>E.coli</u> to produce human protein

1. DNA, polypeptide **2.** nucleic acid **3.** nitrogen-containing base, sugar, phosphoric acid **4.** pyrimidines, purines **5.** ribose, deoxyribose **6.** two, double helix **7.** inside, sugar, phosphate **8.** complementary **9.** hydrogen bonds, guanine, thymine **10.** replication **11.** separate, template, original, new **12.** sequence of bases **13.** mutation

14.(a)   (b)   (c) b

15. T-C-C-A-T-G-C-T-G **16.** (a) coenzymes (b) energy-rich molecules (c) monomers for nucleic acids **17.** c **18.** b **19.** d **20.** c **21.** a **22.** c **23.** d **24.** U-U-A-C-U-C-G-G-A

25. (a) UUC UCA ACU GAU   (b) GUU GCA AGA AAA   (c) AUC CAA CAC CUG
            Phe-Ser-Thr-Asp   Val-Ala-Arg-Lys   Ile-Pro-His-Leu

26. transcription of DNA, translation of RNA **27.** heterogeneous nuclear, introns, exons **28.** mRNA, charged by attaching to an amino acid **29.** smaller, larger **30.** terminal codon **31.** G **32.** D **33.** C **34.** H **35.** A **36.** B **37.** G **38.** C **39.** F

Appendix 1   SOLUTIONS TO THE IN-CHAPTER EXERCISES

Exercise 1-1

1. (a) $48 \text{ mm} \times \dfrac{1 \text{ cm}}{10 \text{ mm}} = 4.8 \text{ cm}$ (b) $3.2 \text{ m} \times \dfrac{1000 \text{ mm}}{1 \text{ m}} = 3200 \text{ mm}$

   (c) $0.03 \text{ km} \times \dfrac{1000 \text{ m}}{1 \text{ km}} = 30 \text{ m}$ (d) $635 \text{ cm} \times \dfrac{1 \text{ in}}{2.54 \text{ cm}} = 250 \text{ in}$

   (e) $7.5 \text{ mi} \times \dfrac{1 \text{ km}}{0.621 \text{ mi}} = 12 \text{ km}$

   (f) $27.3 \text{ m} \times \dfrac{3.28 \text{ ft}}{1 \text{ m}} \times \dfrac{1 \text{ yd}}{3 \text{ ft}} = 29.8 \text{ yd}$

2. (a) $26.2 \text{ mi} \times \dfrac{1 \text{ km}}{0.621 \text{ mi}} = 42.2 \text{ km}$

   (b) Time = 120 min + 8 min + 0.85 min = 128.5 min

     $\dfrac{128.85 \text{ min}}{26.2 \text{ mi}} = 4.92 \text{ mi/min}$

   (c) $\dfrac{128.85 \text{ min}}{42.2 \text{ km}} = 3.05 \text{ km/min}$

Exercise 1-2

1. (a) $253 \ \mu\text{g} \times \dfrac{1 \text{ mg}}{1000 \ \mu\text{g}} = 0.253 \text{ mg}$

   (b) $3.2 \text{ kg} \times \dfrac{1000 \text{ g}}{1 \text{ kg}} = 3200 \text{ g}$

   (c) $0.005 \text{ kg} \times \dfrac{1000 \text{ g}}{1 \text{ kg}} \times \dfrac{1000 \text{ mg}}{1 \text{ g}} = 5000 \text{ mg}$

   (d) $0.34 \text{ kg} \times \dfrac{2.20 \text{ lb}}{1 \text{ kg}} \times \dfrac{16 \text{ oz}}{1 \text{ lb}} = 12 \text{ oz}$

   (e) $681 \text{ g} \times \dfrac{1 \text{ kg}}{1000 \text{ g}} \times \dfrac{2.20 \text{ lb}}{1 \text{ kg}} = 1.50 \text{ lb}$

(f) $30.0 \text{ oz} \times \dfrac{28.3 \text{ g}}{1 \text{ oz}} = 849 \text{ g}$

2. $832 \text{ g} \times \dfrac{1 \text{ lb}}{454 \text{ g}} = 1.83 \text{ lb}$

Exercise 1-3

1. (a) $2.5 \text{ L} \times \dfrac{1000 \text{ mL}}{1 \text{ L}} = 2500 \text{ mL}$

   (b) $345 \text{ mL} \times \dfrac{1 \text{ L}}{1000 \text{ mL}} = 0.345 \text{ L}$

   (c) $25 \text{ mL} = 25 \text{ cc}$

   (d) $343 \text{ gal} \times \dfrac{3.79 \text{ L}}{1 \text{ gal}} = 1300 \text{ L}$

   (e) $5.3 \text{ L} \times \dfrac{1.06 \text{ qt}}{1 \text{ L}} = 5.6 \text{ qt}$

   (f) $945 \text{ mL} \times \dfrac{1 \text{ L}}{1000 \text{ mL}} \times \dfrac{1.06 \text{ qt}}{1 \text{ L}} \times \dfrac{2 \text{ pt}}{1 \text{ qt}} = 2.00 \text{ pt}$

2. 1 liter = \$1 $12 \text{ oz} \times \dfrac{29.6 \text{ mL}}{1 \text{ oz}} \times \dfrac{1 \text{ L}}{1000 \text{ mL}} = .355 \text{ L} \times 3$

   Better buy = 3 12-oz bottles = 1.07 L for \$1

Exercise 1-5

1. $\dfrac{68.9 \text{ g}}{6.10 \text{ cc}} = 11.3 \text{ g/cc}$ 2. $3.6 \text{ g} \times \dfrac{1 \text{ cc}}{1.2 \text{ g}} = 3.0 \text{ cc}$

3. $3.00 \text{ cm} \times 5.50 \text{ cm} \times 2.00 \text{ cm} = 33.0 \text{ cm}^3$ or 33.0 cc

   $33.0 \text{ cc} \times 11.3 \text{ g/cc} = 373 \text{ g}$

Exercise 1-6

   (a) $\dfrac{222 \text{ g}}{200 \text{ cc}} = 1.11 \text{ g/cc}$ (b) $\dfrac{1.11 \text{ g/cc}}{1.00 \text{ g/cc}} = 1.11$

   (c) ethylene glycol

242

Exercise 2-1

1. $1.2 \text{ kg} \times \dfrac{2.20 \text{ lb}}{1 \text{ kg}} \times \dfrac{3500 \text{ kcal}}{1 \text{ lb}} = 9240 \text{ or } 9200 \text{ kcal}$

2. $500 \text{ g} \times (37° - 4°) \times \dfrac{1 \text{ cal}}{\text{g} \, °\text{C}} = 16500 \text{ cal} \times \dfrac{1 \text{ kcal}}{1000 \text{ cal}}$

$$= 16.5 \text{ kcal}$$

Exercise 3-2

silicon

$$
\begin{aligned}
0.9221 \times 28 &= 25.819 \\
0.0470 \times 29 &= 1.363 \\
0.0309 \times 30 &= \underline{0.927} \\
28.109 &= 28.1 \text{ amu}
\end{aligned}
$$

Exercise 5-2

(a)

$$\underset{0}{Mg} \quad + \quad \underset{0}{Br_2} \quad \longrightarrow \quad \underset{2+1-}{MgBr_2}$$

⌐ 2e lost
└ 2 × (1e gained/Br)

(b)

$$\underset{1+1-}{2K\,Cl} + \underset{4+2-}{MnO_2} + \underset{1+\;2-}{H_2SO_4} \longrightarrow \underset{1+2-}{K_2SO_4} + \underset{2+\;2-}{MnSO_4} + \underset{0}{Cl_2} + \underset{1+2-}{H_2O}$$

2e gained/Mn
2 × (1e lost/Cl)

To balance the equation, we must balance the H, SO_4, and O.

$$2KCl + MnO_2 + 2H_2SO_4 \longrightarrow K_2SO_4 + MnSO_4 + Cl_2 + 2H_2O$$

(c)

$$\underset{0}{3Cu} + \underset{1+5+2-}{2H\,N\,O_3} + \underset{1+5+2-}{H\,N\,O_3} \longrightarrow \underset{2+\;5+2-}{3Cu(NO_3)_2} + \underset{2+2-}{2N\,O} + \underset{1+2-}{H_2O}$$

3 × (2e lost/Cu)
2 × (3e gained/N)

To balance the equation, we must balance the H, NO_3, and O.

$$3Cu + 2HNO_3 + 6HNO_3 \longrightarrow 3Cu(NO_3)_2 + 2NO + 4H_2O$$

Exercise 5-3

1. (a) $0.0500 \text{ mol} \times \dfrac{197 \text{ g}}{1 \text{ mol Au}} = 9.85 \text{ g}$

(b) $2.00 \text{ mol} \times \dfrac{65.4 \text{ g}}{1 \text{ mol Zn}} = 131 \text{ g}$

(c) $0.100 \text{ mol} \times \dfrac{32.1 \text{ g}}{1 \text{ mol S}} = 3.21 \text{ g}$

(d) $2.50 \times 10^{20} \text{ atoms} \times \dfrac{1 \text{ mol}}{6.02 \times 10^{23} \text{ atoms}} \times \dfrac{24.3 \text{ g}}{1 \text{ mol Mg}}$

$= 0.0101 \text{ g}$

2. (a) $32.4 \text{ g} \times \dfrac{1 \text{ mol Ag}}{108 \text{ g}} = 0.300 \text{ mol Ag}$

(b) $980 \text{ g} \times \dfrac{1 \text{ mol Si}}{28.1 \text{ g}} = 34.9 \text{ mol Si}$

(c) $0.0202 \text{ g} \times \dfrac{1 \text{ mol Ne}}{20.2 \text{ g}} = 0.00100 \text{ mol} = 1.00 \times 10^{-3} \text{ mol Ne}$

(d) $1.20 \times 10^{25} \text{ atoms} \times \dfrac{1 \text{ mol U}}{6.02 \times 10^{23} \text{ atoms}} = 19.9 \text{ mol U}$

Exercise 5-4

3. $6.02 \times 10^{23} \text{ molecules} \times \dfrac{1 \text{ mol}}{6.02 \times 10^{23} \text{ molecules}} \times \dfrac{254 \text{ g}}{1 \text{ mol}}$

$= 254 \text{ g}$

Exercise 5-5

1.

(a) $0.500 \text{ mol} \times \dfrac{254 \text{ g}}{1 \text{ mol}} = 127 \text{ g}$ (c) $0.0350 \text{ mol} \times \dfrac{123 \text{ g}}{1 \text{ mol}} = 4.30 \text{ g}$

(b) $2.82 \text{ mol} \times \dfrac{239 \text{ g}}{1 \text{ mol}} = 674 \text{ g}$ (d) $4.00 \text{ mol} \times \dfrac{342 \text{ g}}{1 \text{ mol}} = 1370 \text{ g}$

2.

(a) $500 \text{ g} \times \dfrac{1 \text{ mol}}{20.0 \text{ g}} = 25.0 \text{ mol}$ (b) $17.4 \text{ g} \times \dfrac{1 \text{ mol}}{58.1 \text{ g}} = 0.299 \text{ mol}$

(c) $Ca(NO_3)_2 = 40.1 + 2[14.0 + (3 \times 16)] = 164$

$1.76 \text{ g} \times \dfrac{1 \text{ mol}}{164 \text{ g}} = 0.0107 \text{ mol}$

3.

$$1.45 \text{ g} \times \frac{1 \text{ mol}}{58.1 \text{ g}} \times \frac{6.02 \times 10^{23} \text{ molecules}}{1 \text{ mol}} = 1.50 \times 10^{22} \text{ molecules}$$

Exercise 5-7

1.

$$4.18 \text{ g MgCl}_2 \times \frac{1 \text{ mol MgCl}_2}{95.3 \text{ g}} \times \frac{1 \text{ mol MgO}}{1 \text{ mol MgCl}_2} \times \frac{40.3 \text{ g}}{1 \text{ mol MgO}} = 177 \text{ g}$$

2.

$$0.70 \text{ mol H}_2\text{O} \times \frac{2 \text{ mol HCl}}{2 \text{ mol H}_2\text{O}} \times \frac{36.5 \text{ g}}{1 \text{ mol HCl}} = 25.6 \text{ g}$$

$$0.190 \text{ g MgCl}_2 \times \frac{1 \text{ mol MgCl}_2}{95.3 \text{ g}} \times \frac{2 \text{ mol HCl}}{1 \text{ mol MgCl}_2} \times \frac{36.5 \text{ g}}{1 \text{ mol HCl}}$$

$$= 0.146 \text{ g HCl}$$

Exercise 6-1

$$660 \text{ mm Hg} \times \frac{1 \text{ atm}}{760 \text{ mm Hg}} = 0.868 \text{ atm}$$

$$660 \text{ mm Hg} \times \frac{1 \text{ torr}}{1 \text{ mm Hg}} = 660 \text{ torr}$$

Exercise 6-2

1. $760 \text{ torr} \times 3.50 \text{ L} = P_2 \times 4.43 \text{ L}$
   $600 \text{ torr} = P_2$

2. $800 \text{ torr} \times 840 \text{ mL} = 760 \text{ torr} \times V_2$
   $884 \text{ mL} = V_2$

Exercise 6-3

1. $\dfrac{2.0 \text{ L}}{293 \text{ K}} = \dfrac{2.1 \text{ L}}{T_2}$ $T_2 = \dfrac{2.1 \text{ L} \times 293 \text{ K}}{2.0 \text{ L}} = 308 \text{ K} = 35°\text{C}$

2. $\dfrac{1.20 \text{ L}}{293 \text{ K}} = \dfrac{V_2}{310 \text{ K}}$ $V_2 = 1.27 \text{ L}$ $\Delta V = 1.27 \text{ L} - 1.20 \text{ L}$
   $= 0.07 \text{ L} = 70 \text{ mL}$

Exercise 6-4

(a) $P = 1$ atm $T = 273$ K

$$n = 11.4 \text{ g} \times \frac{1 \text{ mol}}{32.0 \text{ g}} = 0.356 \text{ mol} \qquad R = 0.0821 \frac{\text{liter atm}}{\text{mol K}}$$

$$1 \text{ atm} \times V = 0.356 \text{ mol} \times 0.0821 \frac{\text{liter atm}}{\text{mol K}} \times 273 \text{ K}$$

$$V = 7.98 \text{ liters}$$

(b) $P_1 = 1$ atm $= 760$ torr $P_2 = 860$ torr
            $V_1 = 7.98$ L $V_2 = ?$
            $T_1 = 273$ K $T_2 = 293$ K

$$\frac{760 \text{ torr} \times 7.98 \text{ L}}{273 \text{ K}} = \frac{860 \text{ torr} \times V_2}{293 \text{ K}}$$

$$7.57 \text{ L} = V_2$$

(c) $P_1 = 1$ atm $P_2 = 1.50$ atm
            $V_1 = 7.98$ L $V_2 = 3.99$ L
            $T_1 = 273$ K $T_2 = ?$

$$\frac{1 \text{ atm} \times 7.98 \text{ L}}{273 \text{ K}} = \frac{1.50 \text{ atm} \times 3.99 \text{ L}}{T_2}$$

$$T_2 = 205 \text{ K} = -68°C$$

(d) $P_1 = 760$ torr $P_2 = ?$
            $V_1 = 7.98$ L $V_2 = 6.50$ L
            $T_1 = 273$ K $T_2 = 258$ K

$$\frac{760 \text{ torr} \times 7.98 \text{ L}}{273 \text{ K}} = \frac{P_2 \times 6.50 \text{ L}}{258 \text{ K}}$$

$$882 \text{ torr} = P_2$$

Exercise 6-5

$$P_T = P_{oxygen} + P_{water}$$
$$755 \text{ torr} = P_{oxygen} + 23.8 \text{ torr}$$
$$731 \text{ torr} = P_{oxygen}$$

Exercise 7-2

$$5.12 \text{ g} \times \frac{1000 \text{ mg}}{1 \text{ g}} = 5120 \text{ mg}$$

$$80.0 \text{ mg} = 5120 \text{ mg} \times \left(\frac{1}{2} \right)^n$$

$$\frac{1}{64} = \left(\frac{1}{2} \right)^n \qquad n = 6 \text{ half-lives}; \quad 6 \times 13.3 \text{ hr} = 79.8 \text{ hr}$$

Exercise 11-1

1.
(a) $\dfrac{0.200 \text{ mol}}{1000 \text{ mL}} \times \dfrac{106 \text{ g}}{1 \text{ mol } Na_2CO_3} \times 250 \text{ mL} = 5.30 \text{ g}$

(b) $\dfrac{0.75 \text{ mol}}{1 \text{ L}} \times \dfrac{98.0 \text{ g}}{1 \text{ mol } H_3PO_4} \times 1.5 \text{ L} = 110 \text{ g}$

(c) $\dfrac{0.600 \text{ mol}}{1000 \text{ mL}} \times \dfrac{158 \text{ g}}{1 \text{ mol } KMnO_4} \times 150 \text{ mL} = 14.2 \text{ g}$

2. $\dfrac{0.400 \text{ mol}}{1000 \text{ mL}} = \dfrac{0.250 \text{ mol}}{V} \qquad V = 625 \text{ mL}$

3. $320 \text{ mg} \times \dfrac{1 \text{ g}}{1000 \text{ mg}} \times \dfrac{1 \text{ mol } Na^+}{23.0 \text{ g}} \times \dfrac{1000 \text{ mmol}}{1 \text{ mol}} = 13.9 \text{ mmol } Na^+$

$\dfrac{13.9 \text{ mmol } Na^+}{1 \text{ dL}} \times \dfrac{10 \text{ dL}}{1 \text{ L}} = \dfrac{139 \text{ mmol } Na^+}{1 \text{ L}}$

Exercise 11-2

1.
(a) $\dfrac{4.60 \text{ g NaCl}}{100 \text{ mL}} = \dfrac{?}{350 \text{ mL}} \qquad ? = 16.1 \text{ g NaCl}$

(b) $\dfrac{0.220 \text{ g } K_2CO_3}{100 \text{ mL}} = \dfrac{?}{55.0 \text{ mL}} \qquad ? = 0.121 \text{ g } K_2CO_3$

(c) $\dfrac{80 \text{ mg glucose}}{100 \text{ mL}} = \dfrac{?}{25 \text{ mL}} \qquad ? = 20 \text{ mg glucose}$

2. $\dfrac{7.0 \text{ mg uric acid}}{100 \text{ mL}} = \dfrac{5.6 \text{ mg}}{?} \qquad ? = 80 \text{ mL}$

3. $\dfrac{12.0 \text{ g glucose}}{100 \text{ mL}} = \dfrac{?}{275 \text{ mL}} \qquad ? = 33.0 \text{ g glucose}$

Exercise 11-3

$CCl_4: \dfrac{0.40 \text{ } \mu g}{250 \text{ mL}} = \dfrac{?}{1000 \text{ mL}} \qquad ? = 1.6 \text{ } \mu g \qquad 1.6 \text{ ppb}$

Se: $\dfrac{1.3\ \mu g}{250\ mL} = \dfrac{?}{1000\ mL}$ $? = 5.2\ \mu g = 0.0052\ mg$ 0.0054 ppm

Pb: $\dfrac{11\ \mu g}{250\ mL} = \dfrac{?}{1000\ mL}$ $? = 44\ \mu g = 0.044\ mg$ 0.040 ppm

Hg: $\dfrac{0.21\ \mu g}{250\ mL} = \dfrac{?}{1000\ mL}$ $? = 0.84\ \mu g$ 0.84 ppb

Exercise 11-4

2. $140\ mEq \times \dfrac{1\ Eq}{1000\ mEq} \times \dfrac{48.0\ g\ HPO_4^{2-}}{1\ Eq} = 6.72\ g\ HPO_4^{2-}$

3. $\dfrac{94\ mEq}{1\ L} \times \dfrac{1\ Eq}{1000\ mEq} \times \dfrac{35.5\ g\ Cl^-}{1\ Eq} \times \dfrac{1\ mol}{35.5\ g} \times \dfrac{1000\ mmol}{1\ mol}$

$= \dfrac{94\ mmol}{1\ L}$

Exercise 11-5

1.   (a) $\dfrac{2.75\ mol}{1\ L} \times \dfrac{58.5\ g\ NaCl}{1\ mol} \times 1.65\ L = 265\ g\ NaCl$

(b) $1.65\ L \times 2.75\ M = V \times 4.12\ M$ $V = 1.10\ L$

2. $1\ L \times 2.75\ M = 6\ L \times ?$ $? = 0.458\ M$

Exercise 12-3

(a) $[H^+] = \dfrac{1 \times 10^{-14}}{1 \times 10^{-3}} = 1 \times 10^{-11}$ (b) $[H^+] = \dfrac{1 \times 10^{-14}}{5 \times 10^{-6}} = 2 \times 10^{-9}$

(c) $[OH^-] = \dfrac{0.00040\ mol}{200\ mL} \times \dfrac{1000\ mL}{1\ L} = 2 \times 10^{-3}$

$[H^+] = \dfrac{1 \times 10^{-14}}{2 \times 10^{-3}} = 0.5 \times 10^{-11} = 5 \times 10^{-12}$

(d) $[H^+] = \dfrac{1 \times 10^{-14}}{4.0 \times 10^{-4}} = 2.5 \times 10^{-11}$ (e) $[H^+] = \dfrac{1 \times 10^{-14}}{1 \times 10^{-2}} = 1 \times 10^{-12}$

(f) $[OH^-] = \dfrac{8.00 \times 10^{-8}\ mol}{500\ mL} \times \dfrac{1000\ mL}{1\ L} = 1.60 \times 10^{-7}\ M$

$$[H^+] = \frac{1 \times 10^{-14}}{1.60 \times 10^{-7}} = 0.625 \times 10^{-7} = 6.25 \times 10^{-8} \text{ M}$$

Exercise 12-4

1.
   (a) basic, $[H^+] = 1 \times 10^{-11} \text{ M}$, $[OH^-] = \dfrac{1 \times 10^{-14}}{1 \times 10^{-11}} = 1 \times 10^{-3} \text{ M}$

   (b) acidic, $[H^+] = 1 \times 10^{-2} \text{ M}$, $[OH^-] = \dfrac{1 \times 10^{-14}}{1 \times 10^{-2}} = 1 \times 10^{-12} \text{ M}$

   (c) acidic, $[H^+] = 1 \times 10^{-5} \text{ M}$, $[OH^-] = \dfrac{1 \times 10^{-14}}{1 \times 10^{-5}} = 1 \times 10^{-9} \text{ M}$

   (d) basic, $[H^+] = 1 \times 10^{-9} \text{ M}$, $[OH^-] = \dfrac{1 \times 10^{-14}}{1 \times 10^{-9}} = 1 \times 10^{-5} \text{ M}$

2.
   $$[H^+] = \frac{2.0 \times 10^{-9} \text{ mol}}{20 \text{ mL}} \times \frac{1000 \text{ mL}}{1 \text{ L}} = 100 \times 10^{-9} \text{ M} = 1 \times 10^{-7} \text{ M}$$

   pH = 7

Exercise 12-5

2.
   $$\frac{0.120 \text{ Eq}}{1 \text{ L}} \times \frac{36.5 \text{ g HCl}}{1 \text{ Eq}} \times \frac{1 \text{ L}}{1000 \text{ mL}} \times 200 \text{ mL} = 0.876 \text{ g HCl}$$

3.
   $$\frac{0.0500 \text{ Eq}}{1 \text{ L}} \times \frac{37.0 \text{ g Ca(OH)}_2}{1 \text{ Eq}} \times \frac{1 \text{ L}}{1000 \text{ mL}} \times 500 \text{ mL} = 0.925 \text{ g}$$

Exercise 12-6

1.
   $1.5 \text{ L} \times N_a = 0.75 \text{ L} \times 0.10 \text{ N}$
   $N_a = 0.050 \text{ N}$

2.
   $100 \text{ mL} \times N_a = 27.0 \text{ mL} \times 0.10 \text{ N}$
   $N_a = 0.027 \text{ N}$

Exercise 20-1

2.
   $$\frac{1.00 \times 10^2 \text{ molecules}}{1 \text{ min} \quad 1 \text{ molecule E}} \times 1 \times 10^2 \text{ molecules E} \times \frac{60 \text{ min}}{1 \text{ hr}}$$

   $$= \frac{6 \times 10^5 \text{ molecules}}{1 \text{ hour}} \times \frac{1 \text{ mol}}{6 \times 10^{23} \text{ molecules}}$$

   $$= 1 \times 10^{-18} \text{ mol/hour}$$

Exercise 21-1

myristic acid

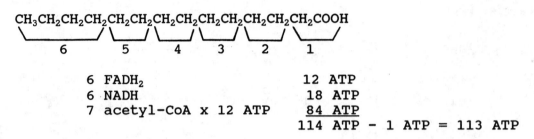

6 FADH$_2$	12 ATP	
6 NADH	18 ATP	
7 acetyl-CoA x 12 ATP	84 ATP	
	114 ATP - 1 ATP = 113 ATP	

Appendix 2   ANSWERS TO THE END-OF-CHAPTER REVIEW PROBLEMS

Chapter 1

1. The mass of an object remains the same wherever the object is located. The weight of an object, which is the measure of the force of gravity on the object, can change with location.

2. (a) Two examples of an element.
   (b) Two examples of a compound.

3. A molecule can be broken down into simpler units (atoms) by ordinary chemical means. An atom cannot.

4. (a) Two examples of foods that are homogeneous.
   (b) Two examples of foods that are heterogeneous.

5. (a) chemical change   (d) physical change
   (b) chemical change   (e) physical change
   (c) physical change   (f) physical change

6. (a) The analytical balance is more precise because both readings were very close. If both scales have been calibrated recently and are in good working order, then the analytical balance is also the more accurate.

7. (a) 3 (b) 2 (c) 5 (d) 3 (e) 6 (f) 3 (g) 1 (h) 3 (i) 7

8. (a) 0.405 kg   (h) 22,400 mL   (o) 0.75 liter
   (b) 1.563 km   (i) 30 cm   (p) 0.005 g
   (c) 160 mm   (j) 0.67 mm   (q) 3.6 dL
   (d) 1500 g   (k) 0.125 g   (r) 46,000 m
   (e) 0.015 liter   (l) 1200 cm   (s) 950 g
   (f) 0.127 m   (m) 0.456 mg   (t) 20 mL
   (g) 70 mg   (n) 1300 mm   (u) 3.61 m

9. (calculator answer is in parentheses)
   (a) 18 ft (17.712)   (l) 3.5 kg (3.4958)
   (b) 3480 ft (3476.8)   (m) 76 cm (76.2)
   (c) 30.0 in   (n) 5900 ft (5904)
   (d) 53 mi (52.785)   (o) 5.00×10^{-4} lb
   (e) 5000 mL   (p) 0.464 gal (0.4643799)
   (f) 16.5 qt (16.536)   (q) 50.0 in
   (g) 7.6 liters (7.58)   (r) 250 g

(h) 170 g (169.8)   (s) 2.4 liters (2.3584905)
(i) 0.231 lb (0.2312775)   (t) 10 pt (9.964)
(j) 109 yd (109.33333)   (u) 0.583 pt
(k) 410 g (410.35)

10. (a) 22°C   (c) -122°C   (e) -45°C
     (b) 194°F   (d) 12°F   (f) 423 K

11. (a) 1883 K, 2930°F   (b) 351.6 K, 173°F   (c) 212.4 K, -77.3°F

12. (a) 1) 0.79 g/cc   2) 0.79   3) 198 g   4) 1.3 mL
     (b) 1) 1.1 g/cc   2) 1.1   3) 275 g   4) 0.91 mL
     (c) 1) 0.98 g/cc   2) 0.98   3) 245 g   4) 1.0 mL

13. (a) 13.6 g/mL or 13.6 g/cc   (b) 3.40 kg

Chapter 2

1. (a) steam
    (b) a car moving at 50 mph
    (c) a football linebacker

2. (a) snow in the mountains
    (b) a pendulum at the top of its swing
    (c) a drawn bow and arrow

3. (a) 3500 kcal   (c) 10 cal   (e) 34.9 cal
    (b) 125,000 cal   (d) 250 cal   (f) 32,300 cal

4. (a) 175 cal, 0.175 kcal   (c) 51,000 cal, 51.0 kcal
    (b) 400 cal, 0.400 kcal   (d) 11,000 cal, 11.0 kcal

5. (a) endothermic   (c) endothermic   (e) exothermic
    (b) exothermic   (d) endothermic

6. yellow light

7. (a) short waves, microwaves, yellow light, ultraviolet
    light, X rays
    (b) X rays, ultraviolet light, yellow light, microwaves,
    short waves

Chapter 3

1.
Subatomic particle	location	charge	relative mass
1) proton	nucleus	1+	1
2) neutron	nucleus	0	1
3) electron	around nucleus	1-	1/1837

2.

	atomic number	symbol	mass number	p	e⁻	n
(a)	3	Li	7	3	3	4
(b)	16	S	32	16	16	16
(c)	11	Na	23	11	11	12
(d)	26	Fe	56	26	26	30
(e)	35	Br	80	35	35	45
(f)	38	Sr	88	38	38	50
(g)	50	Sn	116	50	50	66
(h)	80	Hg	200	80	80	120
(i)	88	Ra	226	88	88	138

3. (a) $Z = 17$, $M = 35$, $^{35}_{17}Cl$, ^{35}Cl, chlorine-35

   (b) $Z = 27$, $M = 60$, $^{60}_{27}Co$, ^{60}Co, cobalt-60

   (c) $Z = 1$, $M = 3$, $^{3}_{1}H$, ^{3}H, hydrogen-3

4.
   zinc-64 $p = 30$ $n = 34$ $e = 30$
   zinc-66 $p = 30$ $n = 36$ $e = 30$
   zinc-67 $p = 30$ $n = 37$ $e = 30$
   zinc-68 $p = 30$ $n = 38$ $e = 30$
   zinc-70 $p = 30$ $n = 40$ $e = 30$

5. The isotopes of zinc differ in the number of neutrons in their nuclei.

6. Br, 80.0   Zn, 65.5

7. The quantum mechanical model of the atom has the protons and neutrons in a small, dense nucleus with the electrons surrounding the nucleus in probability regions called orbitals. Each orbital can hold two electrons spinning in different directions. The farther from the nucleus, the larger the number of orbitals and the greater the energy of the electrons in those orbitals.

8. 4th energy level, 32 electrons; 6th energy level, 72 electrons

9.

		Symbol	Atomic number	Atomic weight	Electron configuration
(a)	lithium	Li	3	6.94	$1s^2\ 2s^1$
(b)	nitrogen	N	7	14.01	$1s^2\ 2s^2\ 2p^3$
(c)	neon	Ne	10	20.18	$1s^2\ 2s^2\ 2p^6$
(d)	magnesium	Mg	12	24.30	$1s^2\ 2s^2\ 2p^6\ 3s^2$
(e)	aluminum	Al	13	26.98	$1s^2\ 2s^2\ 2p^6\ 3s^2\ 3p^1$
(f)	chlorine	Cl	17	35.45	$1s^2\ 2s^2\ 2p^6\ 3s^2\ 3p^5$

10. (a) Be $\quad$ $1s^2\ 2s^2$
    (b) I $\quad$ $1s^2\ 2s^2\ 2p^6\ 3s^2\ 3p^6\ 4s^2\ 3d^{10}\ 4p^6\ 5s^2\ 4d^{10}\ 5p^5$
    (c) Si $\quad$ $1s^2\ 2s^2\ 2p^6\ 3s^2\ 3p^2$
    (d) S $\quad$ $1s^2\ 2s^2\ 2p^6\ 3s^2\ 3p^4$
    (e) Se $\quad$ $1s^2\ 2s^2\ 2p^6\ 3s^2\ 3p^6\ 4s^2\ 3d^{10}\ 4p^4$
    (f) Cu $\quad$ $1s^2\ 2s^2\ 2p^6\ 3s^2\ 3p^6\ 4s^1\ 3d^{10}$
    (g) Mn $\quad$ $1s^2\ 2s^2\ 2p^6\ 3s^2\ 3p^6\ 4s^2\ 3d^5$
    (h) Ca $\quad$ $1s^2\ 2s^2\ 2p^6\ 3s^2\ 3p^6\ 4s^2$
    (i) Ra $\quad$ $1s^2\ 2s^2\ 2p^6\ 3s^2\ 3p^6\ 4s^2\ 3d^{10}\ 4p^6\ 5s^2\ 4d^{10}\ 5p^6\ 6s^2\ 4f^{14}$
    $\quad\quad\quad\quad$ $5d^{10}\ 6p^6\ 7s^2$
    (j) K $\quad$ $1s^2\ 2s^2\ 2p^6\ 3s^2\ 3p^6\ 4s^1$

11. (a) sodium $\quad\quad$ (c) vanadium $\quad\quad$ (e) arsenic
    (b) fluorine $\quad\quad$ (d) iron

12. (a) cation $\quad$ (b) anion $\quad\quad$ (c) anion $\quad\quad$ (d) cation

13. (a) 1 (b) 3 (c) 5 (d) 8 (e) 2 (f) 4 (g) 7 (h) 6

14. (a) metal, representative $\quad\quad$ (e) metal, transition
    (b) nonmetal, representative $\quad$ (f) nonmetal, representative
    (c) metal, transition $\quad\quad\quad$ (g) metalloid, representative
    (d) metalloid, representative (h) metal, representative

15. (a) B, Al, Ga $\quad\quad$ (c) Si, Ga, Cd $\quad\quad$ (e) Cl, O, F
    (b) Te, Sb, Sn $\quad\quad$ (d) Cl, P, As

16. (a) Ca, Mg, Be $\quad\quad$ (b) Te, I, Xe $\quad\quad$ (c) S, Cl, F

17. (a) 1) Na 2) Cl $\quad\quad\quad\quad$ (e) 1) Xe 2) Ne
    (b) 1) C 2) O $\quad\quad\quad\quad\quad$ (f) 1) Sb 2) N
    (c) 1) Rb 2) Li $\quad\quad\quad\quad$ (g) 1) Sr 2) Si
    (d) 1) As 2) F $\quad\quad\quad\quad$ (h) 1) Fe 2) Br

18. (a) Cl $\quad\quad\quad\quad$ (b) Cl $\quad\quad\quad\quad$ (c) Cl

Chapter 4

1. An ionic bond is the force of attraction between two oppositely charged ions, and a covalent bond is formed when two positive nuclei attract and share the same pair of electrons.

2. Ionic bonds are formed between metals and nonmetals, and covalent bonds are formed between nonmetals.

3. Covalent compounds are composed of separate, neutral units called molecules, and ionic compounds are composed of neutral aggregates of ions.

4. (a) Cṡ (b) •Ġe• (c) Cȧ• (d) :N̈e: (e) •Äs• (f) Al̇•

(g) •S̈: (h) •Ï:

5. A single bond is a covalent bond in which two electrons are
shared between two atoms. For example, H:H. A double bond
is a covalent bond in which four electrons are shared
between two atoms. For example, O::C::O. A triple bond is
a covalent bond in which six electrons are shared between
two atoms. For example, N:::N.

6. (a) nonpolar   (d) polar, Cl   (g) polar, F
    (b) polar, Br   (e) polar, O   (h) nonpolar
    (c) polar, N   (f) polar, O

7. Yes, if the polar bonds in a molecule are arranged
symmetrically, then the molecular will be nonpolar.

8. (a) polar   (c) polar   (e) polar
    (b) nonpolar   (d) nonpolar   (f) polar

9. Yes, the hydrogen is attached to the highly electronegative
fluorine atom and would be attracted to a fluorine atom on
another HF molecule.

10. (a) Na: 1+   (e) H : 1+   (i) Fe: 2+
        S : 2-   Br: 5+   P : 5+
                             O : 2-   O : 2-

    (b) Cl: 0   (f) K : 1+   (j) S : 6+
                             Mn: 7+   O : 2-
                             O : 2-

    (c) H : 1+   (g) Na: 1+   (k) S : 4+
        Cl: 3+   H : 1+   O : 2-
        O : 2-   S : 6+
                             O : 2-

    (d) N : 4+   (h) Zn: 2+   (l) Cr: 6+
        O : 2-   N : 5+   O : 2-
                             O : 2-

11. (a) K_2CO_3 (e) $CrCl_3$ (i) $Al_2(SO_3)_3$
    (b) Na_2S (f) $Fe_2(HPO_4)_3$ (j) $Sn(NO_3)_4$
    (c) $Ca(NO_3)_2$ (g) $Cu(C_2H_3O_2)_2$ (k) BeF_2
    (d) SrS   (h) $BaSO_4$ (l) CsBr

12. (a) potassium carbonate   (g) copper(II) acetate
    (b) sodium sulfide   (h) barium sulfate
    (c) calcium nitrate   (i) aluminum sulfite
    (d) strontium sulfide   (j) tin(IV) nitrate
    (e) chromium(III) chloride   (k) beryllium fluoride

(f) iron(III) monohydrogen   (l) cesium bromide
    phosphate

Chapter 5

1. (a) $H_2 + Br_2 \longrightarrow 2HBr$

    (b) $Ca(HCO_3)_2 \longrightarrow CaCO_3 + H_2O + CO_2$

    (c) $2AgNO_3 + Cu \longrightarrow Cu(NO_3)_2 + 2Ag$

    (d) $3H_2 + N_2 \longrightarrow 2NH_3$

    (e) $CH_4 + 4Cl_2 \longrightarrow CCl_4 + 4HCl$

2. (a) $2Na + 2H_2O \longrightarrow 2NaOH + H_2$

    (b) $2KClO_3 \longrightarrow 2KCl + 3O_2$

    (c) $MnO_2 + 4HCl \longrightarrow Cl_2 + MnCl_2 + 2H_2O$

    (d) $C_3H_8 + 5O_2 \longrightarrow 3CO_2 + 4H_2O$

    (e) $4NH_3 + 5O_2 \longrightarrow 4NO + 6H_2O$

    (f) $CO_3^{2-} + 2H^+ \longrightarrow CO_2 + H_2O$

3.

	Group I	Group II	Group III
(a)	40.0 amu	74.6 amu	74.1 amu
(b)	111 amu	98.1 amu	132 amu
(c)	64.1 amu	263 amu	16.0 amu
(d)	164 amu	44.0 amu	36.5 amu
(e)	233 amu	84.0 amu	85.0 amu
(f)	80.9 amu	158 amu	170 amu
(g)	67.8 amu	63.0 amu	62.0 amu
(h)	18.0 amu	160 amu	100 amu

4.

	Group I	Group II	Group III
(a)	6.0 g	313 g	70 g
(b)	278 g	147 g	607 g
(c)	51 g	3.9 g	200 g
(d)	82 g	242 g	0.91 g
(e)	839 g	27 g	53 g
(f)	1.6 g	11.8 g	85 g
(g)	8.48 g	6.3 g	403 g
(h)	0.018 g	9.6 g	0.01 g

5.

	Group I	Group II	Group III
(a)	0.000300 mol	1.20×10^3 mol	5.49 mol
(b)	0.300 mol	0.0749 mol	2.5×10^{-8} mol
(c)	38.7 mol	4.98×10^{-7} mol	9.30 mol
(d)	0.0050 mol	4.50 mol	6.00×10^{-5} mol
(e)	0.0100 mol	8.00×10^{-4} mol	0.760 mol
(f)	6.51×10^{-4} mol	3.16 mol	0.0040 mol
(g)	1.81 mol	1.50×10^{-6} mol	1.30×10^{-3} mol
(h)	3.60×10^{-6} mol	1.99×10^{-2} mol	12.5 mol

6. (a) 0.30 mol $Mg(OH)_2$ 0.10 mol $Mg_3N_2 \times \dfrac{3 \text{ mol } Mg(OH)_2}{1 \text{ mol } Mg_3N_2}$

(b) 14.9 mol $Mg(OH)_2$ $500 \text{ g} \times \dfrac{1 \text{ mol } Mg_3N_2}{101 \text{ g}} \times \dfrac{3 \text{ mol } Mg(OH)_2}{1 \text{ mol } Mg_3N_2}$

9.9 mol NH_3 $500 \text{ g} \times \dfrac{1 \text{ mol } Mg_3N_2}{101 \text{ g}} \times \dfrac{2 \text{ mol } NH_3}{1 \text{ mol } Mg_3N_2}$

(c) 2.0 g Mg_3N_2 0.060 mol $Mg(OH)_2 \times \dfrac{1 \text{ mol } Mg_3N_2}{3 \text{ mol } Mg(OH)_2} \times \dfrac{101 \text{ g}}{1 \text{ mol } Mg_3N_2}$

2.2 g H_2O 0.060 mol $Mg(OH)_2 \times \dfrac{6 \text{ mol } H_2O}{3 \text{ mol } Mg(OH)_2} \times \dfrac{18.0 \text{ g}}{1 \text{ mol } H_2O}$

(d) 30.1 g $Mg(OH)_2$

52.2 g $Mg(OH)_2 \times \dfrac{1 \text{ mol } Mg(OH)_2}{58.3 \text{ g } Mg(OH)_2} \times \dfrac{1 \text{ mol } Mg_3N_2}{3 \text{ mol } Mg(OH)_2} \times \dfrac{101 \text{ g}}{1 \text{ mol } Mg_3N_2}$

(e) 17.3 g $Mg(OH)_2$

$$10.0 \text{ g } Mg_3N_2 \times \frac{1 \text{ mol } Mg_3N_2}{101 \text{ g}} = 0.0990 \text{ mol } Mg_3N_2$$

$$0.0990 \text{ mol } Mg_3N_2 \times \frac{6 \text{ mol } H_2O}{1 \text{ mol } Mg_3N_2} = 0.594 \text{ mol } H_2O$$

$$14.4 \text{ g } H_2O \times \frac{1 \text{ mol } H_2O}{18.0 \text{ g}} = 0.800 \text{ mol } H_2O$$

$$0.0990 \text{ mol } Mg_3N_2 \times \frac{3 \text{ mol } Mg(OH)_2}{1 \text{ mol } Mg_3N_2} \times \frac{58.3 \text{ g}}{1 \text{ mol } Mg(OH)_2} = 17.3 \text{ g}$$

7. (a) H - oxidized   O - reduced
    (b) Br - oxidized   S - reduced
    (c) I - oxidized   N - reduced
    (d) Br - oxidized   Br - reduced

8.   (a)   H - reducing agent   O - oxidizing agent
      (b)   Br - reducing agent   S - oxidizing agent
      (c)   I - reducing agent   N - oxidizing agent
      (d)   Br in NaBr is the reducing agent and Br in $NaBrO_3$ is the oxidizing agent

Chapter 6

1. The particles of an amorphous solid are arranged in a completely random pattern while those in a crystalline solid are arranged in a regular, repeating pattern. Glass is an amorphous solid and quartz is a crystalline solid.

2. 4 kcal $50 \text{ g} \times \dfrac{80 \text{ cal}}{1 \text{ g}} \times \dfrac{1 \text{ kcal}}{1000 \text{ cal}}$

3. (a) motor oil   (b) honey   (c) water at 10°C

4. 27 kcal $50 \text{ g} \times \dfrac{539 \text{ cal}}{1 \text{ g}} \times \dfrac{1 \text{ kcal}}{1000 \text{ cal}}$

5. At 110°C water molecules are in the gaseous state. They have high kinetic energy and are moving very fast in all directions. As the molecules are cooled to 100°C, they move slower and slower. At 100°C the attractions between molecules are strong enough to hold the molecules together when they collide, and a liquid begins to form. When all the molecules are in the liquid state, the temperature will decrease and the molecules will move slower and slower. At 0°C the kinetic energy of the molecules is so low and the attraction between the molecules so strong that the molecules remain in fixed positions in the structure of ice. As the temperature continues to decrease, the vibrations of the molecules in the solid ice become less energetic.

6. Strongest: sugar   weakest: ammonia

7.   (a) 3800 mm Hg   (d) 2100 torr   (g) 14.3 psi
     (b) 0.25 atm   (e) 8400 Pa   (h) 1.17 atm
     (c) 70 mm Hg   (f) 91 kPa

8.   (a) statement of Boyle's law

$P_1V_1 = P_2V_2$, when the temperature and number of moles are constant

     (b) statement of Charles' law

$\dfrac{V_1}{T_1} = \dfrac{V_2}{T_2}$, when pressure and number of moles are constant

(c) statement of Graham's law

$$\frac{\text{Effusion rate (A)}}{\text{Effusion rate (B)}} = \sqrt{\frac{\text{formula weight B}}{\text{formula weight A}}}$$

when the temperature and pressure are held constant

(d) statement of Henry's law

solubility = kP, when the temperature is constant

(e) statement of Dalton's law

$$P_{total} = P_A + P_B + P_C + P_D,\ \text{when the temperature is constant}$$

9. 240 mL

720 torr × 250 mL = 750 torr × V_2

10. 598 mL

$$\frac{505\ \text{mL}}{250\ \text{K}} = \frac{V_2}{296\ \text{K}}$$

11. 22.4 liters

12. 3.3 liters

$$\frac{2.8\ \text{L} \times 909\ \text{mm Hg}}{303\ \text{K}} = \frac{V_2 \times 760\ \text{mm Hg}}{300\ \text{K}}$$

13. V = 115 liters

$$0.750\ \text{atm} \times V = 3.50\ \text{mol} \times 0.0821\ \frac{\text{L atm}}{\text{mol K}} \times 300\ \text{K}$$

14. FW = 64

$$\frac{\text{Effusion rate A}}{4\times\ \text{Effusion rate A}} = \sqrt{\frac{4}{FW_A}}$$

15. The container in which the pressure is 2.5 atm. According to Henry's law, the higher the pressure on a gas, the greater its solubility in water.

16. 684 mm Hg

760 mm Hg = 76 mm Hg + P

17. (a) 738.7 torr

756.2 torr = 17.5 torr + $P_{methane}$

(b) 226 mL

$$\frac{760\ \text{torr} \times V}{273\ \text{K}} = \frac{738.7\ \text{torr} \times 250\ \text{mL}}{293\ \text{K}}$$

18. A gas consists of very small particles randomly moving at very high speeds. Gas particles move in straight lines until they collide elastically with each other or the sides of the container. Gas particles have negligible volume and

there is essentially no interaction between the particles of an ideal gas.

Chapter 7

1. A substance is radioactive when its unstable nucleus emits particles or gives off gamma radiation to become more stable.

2. 6 alpha particles, 4 beta particles

3. Alpha radiation consists of streams of positive helium nuclei which have very little penetrating power. Beta radiation consists of streams of negative electrons which have more penetrating power than alpha particles. Gamma radiation is high energy electromagnetic radiation, which has very high penetrating power.

4. $^{222}_{86}Rn \longrightarrow ^{218}_{84}Po + ^{4}_{2}He$

5. $^{210}_{83}Bi \longrightarrow ^{210}_{84}Po + ^{0}_{-1}e$

6. $^{230}_{90}Th \longrightarrow ^{226}_{88}Ra + ^{4}_{2}He + \gamma$

7. 600 years

8. The third hour. One hour can be thought of as the "half-life" of the money which you have to gamble; after each hour, one-half of your money has "decayed". In radioactive decay, the radioactive element decays to a daughter nucleus which remains. When you gamble and lose, you no longer have that money.

9. Nuclear transmutation is the process by which a different nucleus results from the bombardment of a nucleus by a high-speed particle. An example is the production of carbon-14 by the bombardment of nitrogen-14 with neutrons.

10. (a) $^{4}_{2}He$ (b) $^{10}_{5}B$ (c) $^{1}_{0}n$

11. Atomic fission is the breaking apart of a large, unstable nucleus into smaller, more stable nuclei, and is the process presently used in power plants. Atomic fusion is the combining of small nuclei to form larger, more stable nuclei.

12. There is a relatively small supply of uranium-235, the fuel for current nuclear reactors. Breeder reactors, however,

produce more fuel than they use from several rather abundant isotopes.

13. mill tailings, low-level wastes, transuranic wastes, high-level wastes. Transuranic and high-level wastes pose the most threat to human health.

Chapter 8

1. Ionizing radiation is radiation that can produce highly reactive charged particles when it hits living tissue. Alpha, beta, gamma, and cosmic radiation are all ionizing radiations. Ionizing radiation can do damage to living tissue through direct hits on biologically important molecules or by indirect action, forming ions or free radicals that disrupt the activity of the cell.

2. Damage by direct action is when alpha radiation hits and damages biologically important molecules. Indirect action is caused by beta and gamma radiation and occurs when ions and free radicals are formed in the tissue by the radiation and these particles then do damage to the cell.

3. 9.4×10^{13} disintegrations

4. Beta radiation does its damage to cells by indirect action, forming ions and free radicals. In most cases, cells are able to repair damage from this type of radiation.

5. Alpha particles concentrate their energy in tight, dense paths and do their damage by direct hits on biologically important molecules. This type of damage is difficult for the cell to repair. A cell is more likely to recover from a sublethal dose of beta radiation.

6. You could use a Geiger-Müller counter to locate the area that had been contaminated.

7. 1) Second alternative: the concrete dividing wall acts as shielding from the X rays.
   2) Third alternative: the farther from the source, the lower the intensity of the X rays.
   3) First alternative

8. These tests have doubled this patient's annual background exposure. It is not advisable that this patient have routine dental X rays, because the effects of radiation are cumulative and his exposure way above normal.

9. Images produced by ordinary X rays are two dimensional and often details are masked by overlying tissues. A CT scan

produces cross-sectional images which can be enhanced by color to differentiate specific tissue or body materials.

10. Strontium-90 would be a better choice because it does not give off gamma radiation, which is highly penetrating and could damage nearby normal tissue.

Chapter 9

1. A minimum energy is required to break the glass. When the glass remains intact this energy requirement has not been met. On the occasion that the glass breaks, the minimum energy requirement was met.

2.

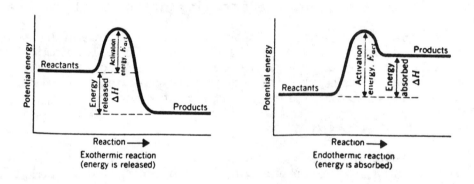

3. (a) exothermic   (c) endothermic
   (b) endothermic   (d) endothermic

4. Add a catalyst, add more reactants, grind up a solid reactant

5. The rate of a chemical reaction is greatly increased if a solid reactant is powdered, and there is a chance of an explosion when finely divided combustible material is being handled.

6. A catalyst is a substance that increases the rate of a chemical reaction without being consumed in the reaction.

7. The drugs deteriorate with time, and placing them in the refrigerator slows the rate of deterioration.

8. (a) The addition of platinum will increase the rate of reaction because the lower activation energy will increase the chances for the molecules to collide with enough energy to react.

(b)

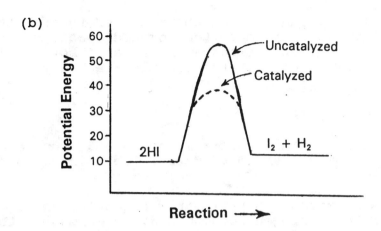

(c)   uncatalyzed: E_{act} = 40.8 kcal   catalyzed: E_{act} = 26 kcal
(d) ΔH = 3 kcal ΔH = 3 kcal

9.   (a) exothermic   (b) 54 kcal   (c) 32 kcal   (d) 86 kcal

10.   The rates off the forward and reverse reactions are equal.

11.   An equilibrium is dynamic because the system is constantly changing, but at equal rates.

12.   (a) $K_c = \dfrac{[NO]^2[O_2]}{[NO_2]^2}$ reactants

      (b) $K_c = \dfrac{[SO_3]^2}{[SO_2]^2[O_2]}$ products

      (c) $K_c = \dfrac{[CO_2][H_2]}{[CO][H_2O]}$ about equal

13.   1) increase the concentration of NO
      2) increase the concentration of Cl_2
      3) increase the pressure

14.   A catalyst has no net effect on an equilibrium system.

15.   statement of Le Chatelier's principle

16.   (a) increase   Increasing the temperature increases the rate of the reverse endothermic reaction.
      (b) decrease   Increasing the pressure will increase the forward reaction thereby lowering the number of molecules.
      (c) decrease   Decreasing the concentration of Cl_2 will increase the forward reaction to replace the lost products.

(d) decrease   Increasing the concentration of the reactant
                HCl will increase the forward reaction.
(e) no change   A catalyst increases the rate of both the
                forward and reverse reactions equally.

Chapter 10

1. Carbon tetrachloride is a nonpolar molecule and ammonia is a
    polar molecule. Ammonia would be more soluble in water than
    carbon tetrachloride.

2. (a) It takes more energy to pull water molecules apart
         because of the hydrogen bonding and, as a result, the
         boiling point is high.
    (b) Hydrogen bonding between water molecules results in the
         open lattice structure of ice, giving ice a low
         density.
    (c) The strong attraction between water molecules makes it
         difficult for them to escape from the liquid and, as a
         result, the heat of vaporization is large.
    (d) Polar molecular substances can form hydrogen bonds with
         water molecules and, as a result, are soluble in water.

3. 10,930 cal $15.2 \text{ g} \times \dfrac{80 \text{ cal}}{\text{g}} = 1220 \text{ cal}$

                        $15.2 \text{ g} \times \dfrac{1 \text{ cal}}{\text{g} \, °C} \times 100 °C = 1520 \text{ cal}$

                        $15.2 \text{ g} \times \dfrac{539 \text{ cal}}{\text{g}} = 8190 \text{ cal}$

4. Final temperature: 32°C $250 \text{ g} \times 5°C \times \dfrac{1 \text{ cal}}{\text{g} \, °C} = 1250 \text{ cal}$

    $250 \text{ g} \times \Delta t \times \dfrac{0.23 \text{ cal}}{\text{g} \, °C} = 1250 \text{ cal} \qquad \Delta t = 22°C$

5. (a) Use the Tyndall effect to distinguish between a
         solution and a colloidal dispersion.
    (b) Let the two solutions stand, the suspension will settle
         out while the colloidal dispersion will not.

6. The solvent is the dissolving medium. The solute is the
    substance being dissolved, and the solution is the mixture
    of the solute and the solvent.

7. Description of the steps that occur when NaCl is dissolved
    in water.

8. (a) no, nonelectrolyte   (c) yes, electrolyte
    (b) yes, electrolyte

9. The solubility of a gas in a liquid decreases with an
    increase in temperature. Hence, more carbon dioxide will be
    dissolved in the soft drink when it is cold than when it is
    at room temperature.

10. (a) no   (b) yes   (c) yes   (d) no   (e) yes   (f) no

11. $H^+_{(aq)} + OH^-_{(aq)} \longrightarrow H_2O_{(1)}$

Chapter 11

1. A dilute solution contains only a few solute particles while
    a concentrated solution contains many solute particles.
    More specifically, a saturated solution contains all of the
    solute particles the solvent can usually hold at a certain
    temperature. If the temperature decreases in a saturated
    solution but none of the solute particles crystalize, then
    the solution is supersaturated and contains more solvent
    particles than normal at that temperature.

2. There will be no net change in the amount of dissolved
    solute but since an equilibrium exists, some of the added
    sugar will go into solution as other sugar molecules
    precipitate out of solution.

3. 0.20 M H_2SO_4 $\dfrac{4.9 \text{ g } H_2SO_4}{250 \text{ mL}} \times \dfrac{1 \text{ mol}}{98 \text{ g}} \times \dfrac{1000 \text{ mL}}{1 \text{ L}}$

4. 4.44 g hexachlorophene $148 \text{ mL} \times \dfrac{3.00 \text{ g}}{100 \text{ mL}}$

5. 2500 mL $2.5 \text{ g} \times \dfrac{100 \text{ mL}}{0.10 \text{ g } Na_2CO_3}$

6. Dissolve 0.62 mg in enough $25 \text{ mL} \times \dfrac{2.5 \text{ mg}}{100 \text{ mL}}$
    water to make 25 mL of solution

7. (a) Dissolve 0.40 g of NaOH in enough water to make 50 mL
         of solution.
    (b) Dissolve 3.9 g of Na_2SO_4 in enough water to make 250 mL
         of solution.
    (c) Dissolve 25 g of KOH in enough water to make 0.50 liter
         of solution.
    (d) Dissolve 2.7 g of NH_4Cl in enough water to make 0.10
         liters of solution.
    (e) Dissolve 10.3 g of glucose in enough water to make 125
         mL of solution.

(f) Dissolve 49.5 mg of lactic acid in enough water to make 225 mL of solution

(g) Dissolve 12 g of HCl in enough water to make 55 mL of solution.

(h) Dissolve 0.35 g of K_2HPO_4 in enough water to make 10 mL of solution.

(i) Dissolve 0.33 g of $(NH_4)_2SO_4$ in enough water to make 0.25 liters of solution.

(j) Dissolve 0.291 g of KCl in enough water to make 35 mL of solution.

8. 1.7 ppb

$$\frac{0.26\ \mu g\ Cd}{150\ mL} \times \frac{1000\ mL}{1\ L}$$

9. 40 ppm

$$\frac{4\ mg\ Pb}{100\ mL} = \frac{1000\ mL}{1\ L}$$

10. (a) 1) 4.0×10^{-6} M

$$\frac{11\ \mu g}{10\ mL} = \frac{1\ mol}{272 \times 10^6\ \mu g} = \frac{1000\ mL}{1\ L}\ H_2O$$

   2) 0.11 mg% $11\ \mu g \times \dfrac{1\ mg}{1 \times 10^3\ \mu g}$; $\dfrac{1.1 \times 10^{-2}\ mg}{10\ mL} = \dfrac{?\ mg}{100\ mL}$

   3) 1.1 ppm

$$\frac{0.11\ mg}{100\ mL} = \frac{?\ mg}{1000\ mL}$$

   (b) 1) 4.8×10^{-4} M   2) 2.0 mg%   3) 20 ppm
   (c) 1) 6.0×10^{-3} M   2) 39 mg%   3) 390 ppm
   (d) 1) 1.2×10^{-5} M   2) 0.13 mg%   3) 1.3 ppm

11. (a) 31.7 g   (b) 79.9 g   (c) 68.7 g   (d) 30.0 g

12. Yes, the blood potassium level is 2.0 mEq/liter

$$\frac{3.9\ mg\ K^+}{50\ mL} \times \frac{1\ mEq}{39.1\ mg\ K^+} \times \frac{1000\ mL}{1\ L}$$

13. 0.24 mg Mg^{2+}

$$10\ mL \times \frac{2.0\ mEq}{1000\ mL} \times \frac{12\ mg}{1\ mEq\ Mg^{2+}}$$

14. (a) 1.2 M H_2SO_4 (c) 0.6% KCl
    (b) 0.4 ppm Cd^{2+} (d) 6.4 mg% urea

15. (a) 200 mL of stock solution plus enough water to make 250 mL of solution.
    (b) 45 mL of stock solution plus enough water to make 150 mL of solution.
    (c) 75 mL of stock solution plus enough water to make 750 mL of solution.

(d) 45 mL of stock solution plus enough water to make 55 mL of solution.

(e) 260 mL of stock solution plus enough water to make 650 mL of solution.

16. Salt lowers the freezing point of water, thus melting the ice.

17. When celery has gone limp, it has lost water from its cells. Water is hypotonic to the cells in the celery. When the celery is placed in water, water molecules will move into the cells of the celery, restoring its crispness.

18. (a) 1   (b) 2   (c) 1

19. 0.3 osmols/liter $\dfrac{0.9 \text{ g NaCl}}{100 \text{ mL}} \times \dfrac{1 \text{ mol}}{58.5 \text{ g}} \times \dfrac{1000 \text{ mL}}{1 \text{ L}} = 0.15 \text{ M}$

20. (a) crenation   (b) no change   (c) hemolysis

21. (a) hypertonic   (b) isotonic   (c) hypotonic

22. Schematic diagram and description of an artificial kidney as described in Section 11.11.

Chapter 12

1. Statement of the Bronsted-Lowry definition of an acid and base.

2. A polyprotic acid can donate more than one proton in a reaction with a base. H_2SO_4, H_3PO_4

3. (a) $\underset{\text{acid}_1}{HI}$ + $\underset{\text{base}_2}{H_2O}$ $\rightleftharpoons$ $\underset{\text{acid}_2}{H_3O^+}$ + $\underset{\text{base}_1}{I^-}$

(b) $\underset{\text{base}_1}{CO_3^{2-}}$ + $\underset{\text{acid}_2}{H_2O}$ $\rightleftharpoons$ $\underset{\text{base}_2}{OH^-}$ + $\underset{\text{acid}_1}{HCO_3^-}$

(c) $\underset{\text{acid}_1}{CH_3COOH}$ + $\underset{\text{base}_2}{H_2O}$ $\rightleftharpoons$ $\underset{\text{acid}_2}{H_3O^+}$ + $\underset{\text{base}_1}{CH_3COO^-}$

(d) $\underset{\text{acid}_1}{HF}$ + $\underset{\text{base}_2}{NH_3}$ $\rightleftharpoons$ $\underset{\text{acid}_2}{NH_4^+}$ + $\underset{\text{base}_1}{F^-}$

(e) $\underset{\text{base}_1}{O^{2-}}$ + $\underset{\text{acid}_2}{H_2O}$ $\rightleftharpoons$ $\underset{\text{base}_2}{OH^-}$ + $\underset{\text{acid}_1}{OH^-}$

(f) $\underset{\text{base}_1}{NO_2^-}$ + $\underset{\text{acid}_2}{N_2H_5^+}$ $\rightleftharpoons$ $\underset{\text{acid}_1}{HNO_2}$ + $\underset{\text{base}_2}{N_2H_4}$

(g) HCl   +   NH$_2$OH   NH$_3$OH$^+$ +   Cl$^-$
     acid$_1$ base$_2$ acid$_2$ base$_1$

4. (a) H$_3$O$^+$ (b) H$_2$CO$_3$ (c) H$_2$PO$_4^-$ (d) HCl

5. (a) OH$^-$ (b) PO$_4^{3-}$ (c) NH$_3$ (d) CO$_3^{2-}$

6. (a) hydroiodic acid   (c) chlorous acid
    (b) sulfurous acid   (d) chloric acid

7. (a) potassium iodide   (c) potassium chlorite
    (b) potassium sulfite   (d) potassium chlorate

8. (a) NH$_4$Cl + NaOH $\longrightarrow$ H$_2$O + NH$_3$ + NaCl

    (b) NH$_4^+$ + OH$^-$ $\longrightarrow$ H$_2$O + NH$_3$

The odor of ammonia would be greater after the addition of
sodium chloride because ammonia is a product of the reaction
between ammonium chloride and sodium hydroxide.

9. H$_3$AsO$_4$ + 3NaOH $\longrightarrow$ 3H$_2$O + Na$_3$AsO$_4$

       3H$^+$ + 3OH$^-$ $\longrightarrow$ 3H$_2$O

10.   [H$^+$]   [OH$^-$]
    (a) 1 $\times$ 10^{-1} 1 $\times$ 10^{-13}
    (b) 1 $\times$ 10^{-6} 1 $\times$ 10^{-8}
    (c) 1 $\times$ 10^{-12} 1 $\times$ 10^{-2}

11. 12, 9.5, 8.4, 7.4   7   6.3, 5.5, 4, 1.4
          basic   neutral   acidic

12. (a) 2   (b) 8   (c) 10

13. (a) acidic   (b) basic   (c) basic

14. (a) 26.0 g   (c) 29.2 g   (e) 80.9 g
     (b) 50.0 g   (d) 47.3 g   (f) 23.9 g

15. (a) 1.73 Eq Al(OH)$_3$ (b) 0.12 Eq H$_2$SO$_4$ (c) 0.20 Eq NaOH

16. Solution 1:

     (a) 0.010 M HCl $\dfrac{0.0365 \text{ g HCl}}{100 \text{ mL}} \times \dfrac{1 \text{ mol}}{36.5 \text{ g}} \times \dfrac{1000 \text{ mL}}{1 \text{ L}}$

     (b) 0.010 N HCl $\dfrac{0.0365 \text{ g HCl}}{100 \text{ mL}} \times \dfrac{1 \text{ Eq}}{36.5 \text{ g}} \times \dfrac{1000 \text{ mL}}{1 \text{ L}}$

     (c) [H$^+$] = 0.0100 = 1 $\times$ 10^{-2} (d) pH = 2

Solution 2:
(a) 0.0500 M H_2SO_4
(b) 0.100 N H_2SO_4

(c) $[H^+] = 1 \times 10^{-1}$
(d) pH = 1

Solution 3:
(a) 5.00×10^{-6} M $Ca(OH)_2$
(b) 1.00×10^{-5} M $Ca(OH)_2$

(c) $[H^+] = 1 \times 10^{-9}$
(d) pH = 9

Solution 4:
(a) 0.100 M NaOH
(b) 0.100 N NaOH

(c) $[H^+] = 1 \times 10^{-13}$
(d) pH = 13

17. 8.00 g NaOH

$$7.30 \text{ g} \times \frac{1 \text{ Eq HCl}}{36.5 \text{ g}} = 0.200 \text{ Eq HCl}$$

$$0.200 \text{ Eq} \times \frac{40.0 \text{ g}}{1 \text{ Eq NaOH}} = 8.00 \text{ g NaOH}$$

18. 18.4 g $NaHCO_3$

$$36.5 \text{ mL} \times \frac{6.00 \text{ Eq}}{1000 \text{ mL}} = 0.219 \text{ Eq } H_2SO_4$$

$$0.219 \text{ Eq} \times \frac{84.0 \text{ g}}{1 \text{ Eq NaHCO}_3} = 18.4 \text{ g NaHCO}_3$$

19. 400 mL

0.85 M H_2SO_4 = 1.70 N H_2SO_4

3.40 M NaOH = 3.40 N NaOH

V × 1.70 N H_2SO_4 = 200 mL × 3.40 N NaOH

20. The following equilibrium exists in the buffer system:

$$H_2CO_3 \rightleftharpoons HCO_3^- + H^+$$

(a) When a acid is added to the system, the equilibrium will shift to the left, thus lowering the H^+ concentration.
(b) When a base is added to the system, the equilibrium will shift to the right, to replace the H^+ that reacts with the base.

21. (a) phosphate buffer system   (b) carbonic acid/bicarbonate

22. (a) In acidosis the rate of breathing will increase to expel excess carbon dioxide, and the kidneys will increase the excretion of hydrogen ion while lowering the excretion of bicarbonate.
(b) In alkalosis, the rate of breathing will slow down to retain carbon dioxide. The rate of excretion of hydrogen ions will be reduced while the rate of

excretion of basic components of the blood by the
kidneys will be increased.

Chapter 13

1. Molecules found in living organisms are mainly long chains,
    branched chains, or rings which have carbon atoms as the
    axis or backbone of the molecule. Carbon is able to form
    strong stable bonds with up to four other carbon atoms, and
    also with atoms of elements such as oxygen, hydrogen,
    sulfur, phosphorus, and nitrogen.

2. (a) Hybrid orbitals are formed when atomic orbitals mix
         together. These hybrid orbitals allow for greater
         overlap and the formation of stronger covalent bonds.

    (b) sp, sp^2, sp^3

    (c) sp^3 hybrid orbitals

3. Definition of organic chemistry.

4. The molecular formula of a compound indicates only the
    number and type of atoms in a molecule. The structural
    formula shows the geometric arrangement of these atoms.

5. $CH_3CH_2CH_2CH_2CH_2CH_3$

    $CH_3CH_2CHCH_2CH_3$
    $\qquad\quad |$
    $\qquad\quad CH_3$

    $\qquad CH_3$
    $\qquad |$
    $CH_3-C-CH_2CH_3$
    $\qquad |$
    $\qquad CH_3$

    $CH_3CHCH_2CH_2CH_3$
    $\quad\ |$
    $\quad\ CH_3$

    $\qquad CH_3\ \ CH_3$
    $\qquad |\qquad |$
    $CH_3 - CH - CH - CH_3$

6. (a) propane
    (b) nonane
    (c) methylbutane
    (d) 2,4-dimethylpentane
    (e) methylpropane

    (f) 3-methyl-4-ethylheptane
    (g) 3-isopropylhexane
    (h) 4-methylheptane
    (i) 4-propyl-4-isopropyloctane

7. (a) $CH_3(CH_2)_3-CH-(CH_2)_4CH_3$
         $\qquad\qquad\quad |$
         $\qquad\qquad\quad CH_2CH_3$

    (d) $\qquad\ Cl\ \ Cl$
         $\qquad\ |\ \ \ |$
         $Cl-CH-CH-CH_2CH_3$

(b)

$$CH_3 \overset{\overset{\displaystyle CH_3}{|}}{\underset{\underset{\displaystyle CH_3}{|}}{C}} CH_3$$

(e)

$$CH_3\overset{\overset{\displaystyle I}{|}}{C}CH_2\overset{\overset{\displaystyle CH_3}{|}}{C}HCH_2\overset{\overset{\displaystyle CH_3}{|}}{C}HCH_3$$
$$\underset{\underset{\displaystyle CH_3}{|}}{}$$

(c) $CH_3(CH_2)_2-CH-(CH_2)_3CH_3$
$$\qquad\qquad\overset{|}{CH_3-CH-CH_3}$$

(f)

$$CH_3\overset{\overset{\displaystyle CH_3}{|}}{C}HCH_2\overset{\overset{\displaystyle CH_2CH_2CH_2CH_3}{|}}{C}HCH(CH_2)_3CH_3$$
$$\qquad\qquad\overset{|}{CH_3}$$

8. (a) 2-nitropropane
    (b) 3-ethyl-3-methylpentane
    (c) 2-iodo-3-methylbutane
    (d) 3-chloro-2,2-dimethyl-3-ethylpentane

9. (a)

$$Br-\overset{\overset{\displaystyle Br}{|}}{C}H-\overset{\overset{\displaystyle Br}{|}}{C}H-Br$$

(d)

$$CH_3CH_2-\overset{\overset{\displaystyle CH_2CH_3}{|}}{\underset{\underset{\displaystyle CH_2CH_3}{|}}{C}}-CH_2CH_3$$

(b)

$$CH_3-\overset{\overset{\displaystyle NH_2}{|}}{C}H-(CH_2)_4CH_3$$

(e) $NH_2CH_2CH_2CH_2CH_2NH_2$

(c)

$$CH_3-\overset{\overset{\displaystyle CH_3}{|}}{C}H-\overset{\overset{\displaystyle Cl}{|}}{\underset{\underset{\displaystyle CH_3}{|}}{C}}-CH_2CH_2CH_3$$

(f) $FCH_2(CH_2)_2CH(CH_2)_3CH_3$
$$\qquad\qquad\qquad\overset{|}{CH_2}\overset{}{C}HCH_3$$
$$\qquad\qquad\qquad\qquad\overset{|}{CH_3}$$

10. (a) the same   (i) the same
    (b) constitutional isomers   (j) unrelated
    (c) unrelated   (k) unrelated
    (d) the same   (l) constitutional isomers
    (e) the same   (m) the same
    (f) constitutional isomers   (n) unrelated
    (g) constitutional isomers   (o) unrelated
    (h) constitutional isomers   (p) the same

11. (a) $C_3H_8 + 5O_2 \longrightarrow 3CO_2 + 4H_2O + energy$

    (b) $2C_8H_{18} + 25O_2 \longrightarrow 16CO_2 + 18H_2O + energy$

    (c) $2C_6H_{14} + 19O_2 \longrightarrow 12CO_2 + 14H_2O + energy$

12. (a) $CH_4 + Br_2 \longrightarrow CH_3Br + HBr$

    (b) $C_2H_6 + HNO_3 \longrightarrow C_2H_5NO_2 + H_2O$

(c) $C_3H_8 + Br_2 \longrightarrow C_3H_7Br + HBr$ ($CH_3CH_2CH_2Br$, $CH_3CHBrCH_3$)

Chapter 14

1. (a) $CH_3CHCH=CHCH_2CH_2CH_3$
             |
            CH_3

   (e)
            CH_3
             |
      $CH_3CH=CHCH_2CCH_3$
             |
            CH_3

   (b)
            CH_3 CH_3
             | |
      $CH_3 - C = C - CH_3$

   (c) $CH_2=CH-CH=CH-CH=CHCH_2CH_3$

   (f)
            CH_3 $CH_2CH_2CH_3$
             | |
      $CH_2=C-CHCHCHCH_2CH_3$
             | |
            CH_3 CH_3

   (d)
                  Br   Br
                   | |
      $CH_2=CH-C - C-CH_2-CH_3$
                   | |
                  Br   Br

2. (a) one sigma and one pi bond
   (b) one sigma and two pi bonds

3. (a) $CH_2=CHCH=C=CHCH_3$
        1   2 3 4 5 6

       Carbon 1,2,3,5: 3 sigma and 1 pi
       Carbon 4: 2 sigma and 2 pi
       Carbon 6: 4 sigma

   (b) $CH\equiv C-CH=CH-CH_3$
        1 2 3   4   5

       Carbon 1,2: 2 sigma and 2 pi
       Carbon 3,4: 3 sigma and 1 pi
       Carbon 5: 4 sigma

4. (a)
       CH_3 CH_2CH_3
          \   /
           C=C
          /   \
        H   H

       CH_3 H
          \   /
           C=C
          /   \
        H CH_2CH_3

   (b)
       Cl   Cl
          \   /
           C=C
          /   \
        H   H

       Cl   H
          \   /
           C=C
          /   \
        H   Cl

(c)

```
            H                          H
            |                          |
 H          C          H    H          C          H
 |         / \         |    |         / \         |
 C -- H   /   \   H -- C    C -- H   /   \   Cl -- C
 |       C     C       |    |       C     C       |
 H       |     |       H    H       |     |       H
         Cl    Cl                   Cl    H
```

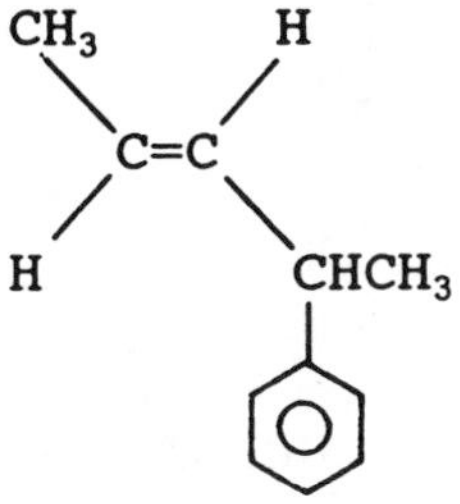

5. **(a)**

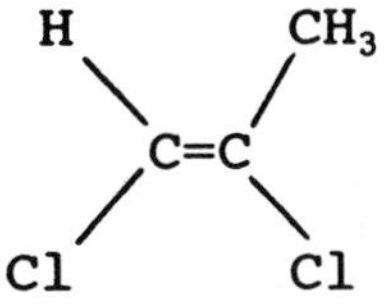

```
  H        CH3
   \      /
    C == C
   /      \
  Cl       Cl
```

(b)

```
  CH3       H
   \       /
    C  ==  C
   /       \
  H         CHCH3
              |
          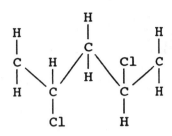
```

(c)

```
  CH3CH2        CH2CH3
       \       /
        C  ==  C
       /       \
      H          H
```

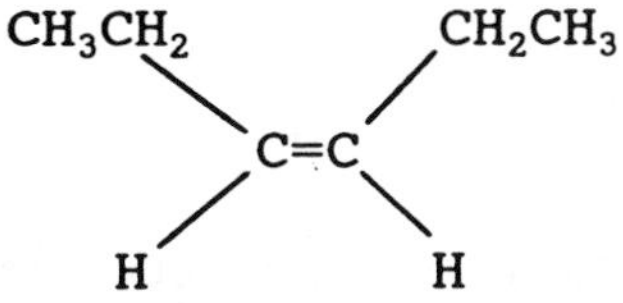

(d)

```
              CH3CH2   CH2CH2CH2CH3
                 |       |
  CH3CH2CH2 — C == C — CHCH2CH2CH2CH3
                 |
              CH2CH3
```

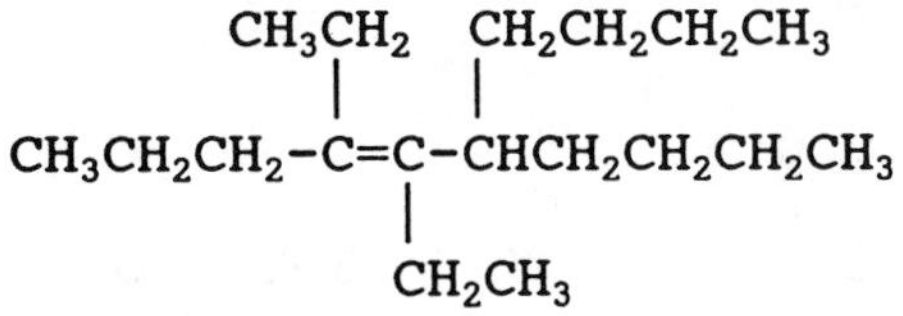

6. **(a)** $C_4H_8 + 6O_2 \longrightarrow 4CO_2 + 4H_2O$

(b) $CH_3CH=CHCH_2CH_3 + H_2 \longrightarrow CH_3CH_2CH_2CH_2CH_3$

7.

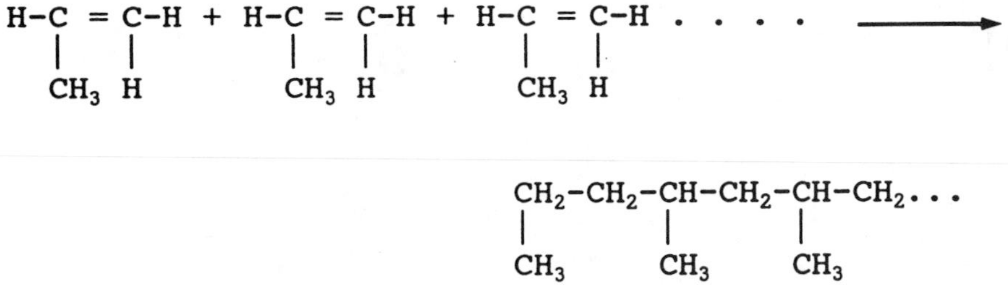

```
H–C = C–H  +  H–C = C–H  +  H–C = C–H  . . . .  ⟶
   |   |         |   |         |   |
   CH3 H         CH3 H         CH3 H
```

```
            CH2–CH2–CH–CH2–CH–CH2. . .
                     |       |       |
                     CH3     CH3     CH3
```

8. (a) $CH_3C{\equiv}CCH_2CH_2CH_3$

(b)
$$CH_3C{\equiv}C{-}\overset{\displaystyle CH_2CH_3}{\underset{\displaystyle CH_2CH_3}{C}}{-}CHCH_2CH_3$$

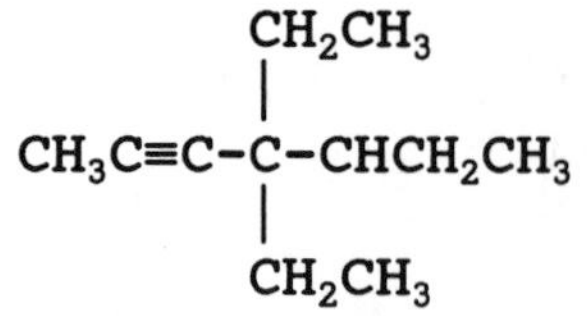

(c)
$$CH{\equiv}C{-}\overset{\displaystyle CH_3}{\underset{\displaystyle CH_3}{C}}{-}\overset{}{\underset{\displaystyle CH_3}{CH}}{-}CH_3$$

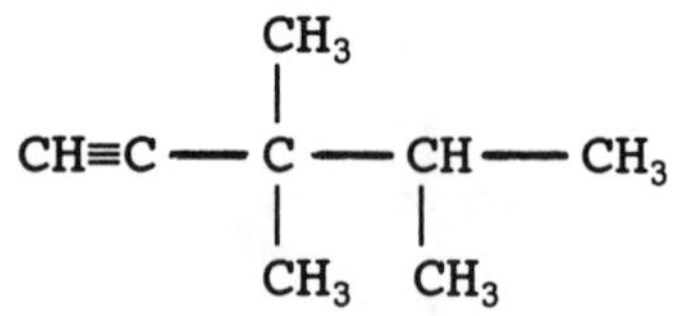

(d)
$$CH_3C{\equiv}C\underset{\displaystyle CH_3}{CHCH_3}$$

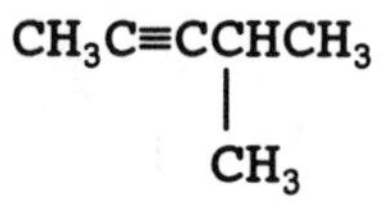

(e)
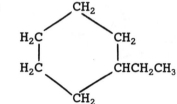

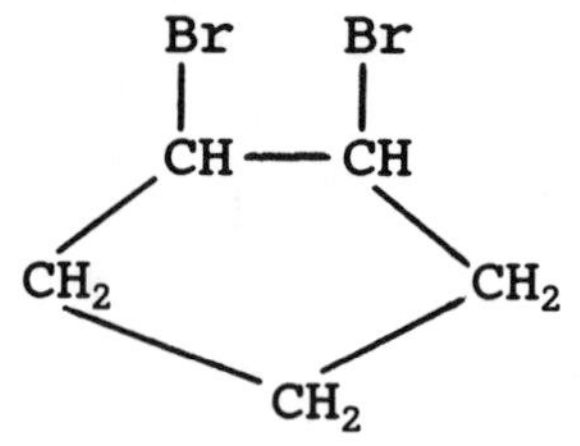

(f)

9. Aromatic compounds have a resonance hybrid structure in the ring that gives the ring extra stability. These ring structures will only undergo substitution reactions and not addition reactions.

10. b and c are aromatic compounds

11. CH_2Br_2 bromophenylmethane CH_3 3-bromotoluene

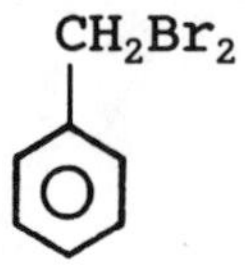

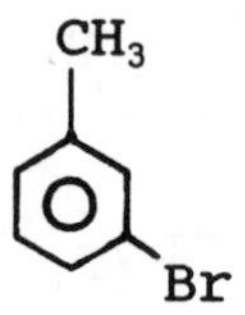

CH_3 2-bromotoluene CH_3 4-bromotoluene

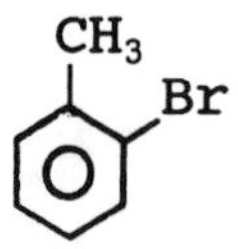

12. (a) NH_2 CH_3 (d) $CH_3C{=}CHCH_3$

274

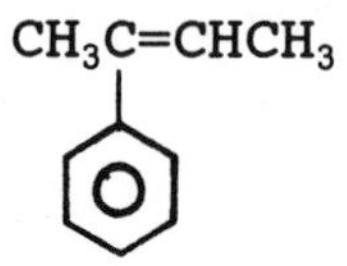

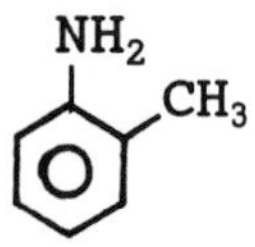

(b)

(e)

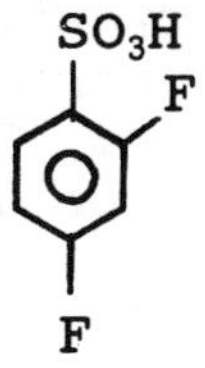

(c)

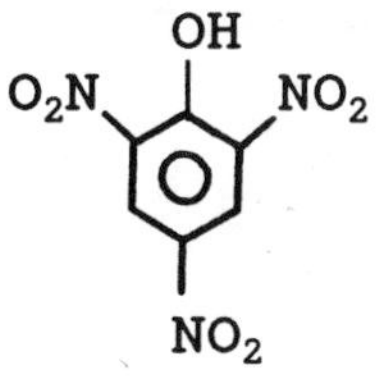

13. (a)

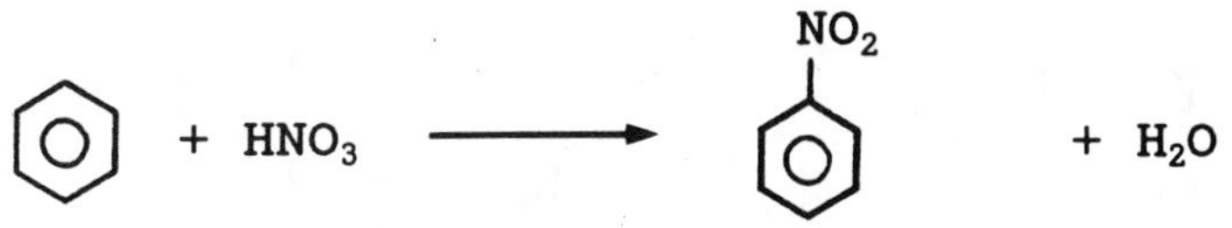

(b)

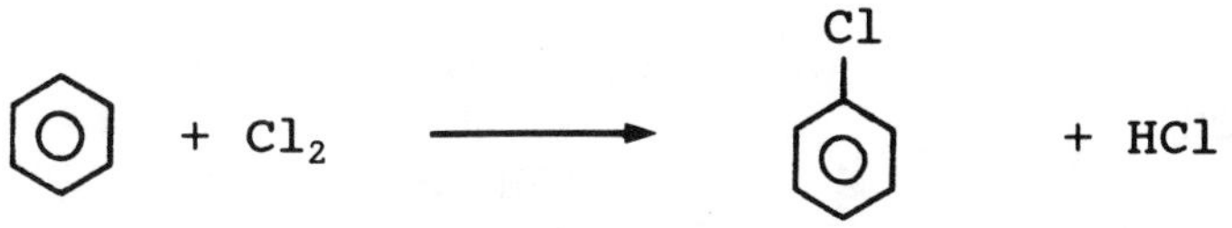

Chapter 15

1. (a) 2-butanol, secondary
   (b) benzyl alcohol, primary
   (c) 2-methyl-2-butanol, tertiary
   (d) 1-propanol, primary
   (e) 3,3-dimethyl-2-butanol, secondary
   (f) 2-phenyl-2-propanol, tertiary

2. (a) $CH_2ClCH_2CH_2OH$

   (b) 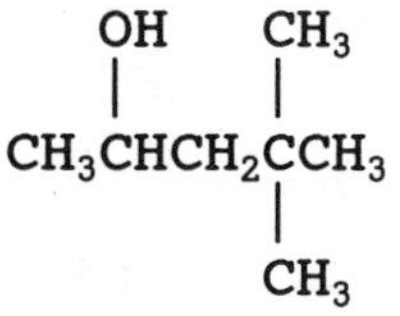
   $$CH_3CHCH_2CCH_3$$
   with OH on second carbon and CH_3 groups on fourth carbon (2-methyl...)

   (c) $HOCH_2CH_2CHOHCH_2CH_3$

   (d) 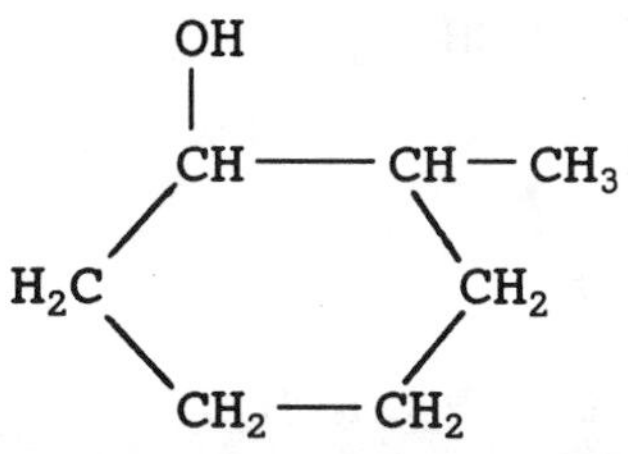 (ring structure with OH)

3. The hydroxyl group creates a polar spot on the otherwise nonpolar hydrocarbon chain, and can form a hydrogen bond with a water molecule. Therefore, increasing the number of hydroxyl groups on a molecule increases its solubility in water.

4. Ethanol has a higher boiling point than propane and methyl ether because of the hydrogen bonding that can form between the molecules of ethanol.

5. You would add glycerol to the cherry filling as a sweetener and as a protection against drying out.

6. (a) $CH_3CH_2CH=CH_2$ (b) $CH_3CH_2CH=CHCH_2CH_3$

7. (a)
$$\underset{CH_3CH_2CH}{\overset{O}{\parallel}}$$

(e) H_2C——CH_2 | | H_2C——CH_2 $\rangle C=O$

(b)
$$\underset{CH_3CCH_3}{\overset{O}{\parallel}}$$

(f)
$$\underset{CH_3CH}{\overset{O}{\parallel}}$$

(c)
$$\underset{CH_3CHCH_2CH}{\overset{CH_3 \quad O}{\overset{|}{}\overset{\parallel}{}}}$$

(g)
$$\underset{CH_3CH_2CH_2CH_2CH}{\overset{O}{\parallel}}$$

(d)
$$\underset{\underset{CH_3}{|}}{\underset{CH_3CHCCH_3}{\overset{O}{\parallel}}}$$

(h)
$$\underset{CH_3CH_2CH_2CH_2CCH_3}{\overset{O}{\parallel}}$$

8. (a) $CH_3CH_2CH_2OCH_2CH_2CH_3$

(b)
$$\underset{CH_3CHCH_2OCH_3}{\overset{CH_3}{\overset{|}{}}}$$

(c) ⟨O⟩–O–CH₂CH₃

9. (a) methyl butanoate (b) 2-ethylbutanoic acid

10. (a)
$$\underset{CH_3CH_2CHCH_3}{\overset{OH}{\overset{|}{}}}$$

(d)
$$\underset{CH_3CH_2CH_2CHCH_3}{\overset{OH}{\overset{|}{}}}$$

(b) $CH_3CH_2CH_2CH_2OH$

(e)
$$\underset{CH_3CHCH_3}{\overset{OH}{\overset{|}{}}}$$

(c) ⟨O⟩–CH₂OH

(f) $CH_3CH_2CH_2CH_2CH_2CH_2OH$

11. (a)

$$CH_3\overset{\overset{\displaystyle O}{\|}}{C}OH$$

(c)

$$CH_3CH_2CH_2CH_2CH_2\overset{\overset{\displaystyle O}{\|}}{C}OH$$

(b)

$$CH_3CH_2CH_2\overset{\overset{\displaystyle O}{\|}}{C}OH$$

(d)

12. (a)

$$CH_3-\underset{\underset{\displaystyle OCH_3}{|}}{\overset{\overset{\displaystyle OH}{|}}{C}}-H$$

(b)

$$CH_3-\underset{\underset{\displaystyle OCH_2CH_3}{|}}{\overset{\overset{\displaystyle OH}{|}}{C}}-CH_3$$

(c)

$$CH_3CH_2-\underset{\underset{\displaystyle OCH_3}{|}}{\overset{\overset{\displaystyle OH}{|}}{C}}-CH_3$$

(d)

$$H-\underset{\underset{\displaystyle OCH_2CH_3}{|}}{\overset{\overset{\displaystyle OH}{|}}{C}}-H$$

(e)

$$CH_3CH_2-\underset{\underset{\displaystyle OCH_3}{|}}{\overset{\overset{\displaystyle OH}{|}}{C}}-H$$

(f)

$$CH_3-\underset{\underset{\displaystyle OCH_2CH_2CH_3}{|}}{\overset{\overset{\displaystyle OH}{|}}{C}}-CH_3$$

13. (a) heptanoic acid

(c)

$$CH_3-\underset{\underset{\displaystyle Cl}{|}}{\overset{\overset{\displaystyle CH_3}{|}}{C}}-\overset{\overset{\displaystyle CH_3}{|}}{CH}-\overset{\overset{\displaystyle O}{\|}}{C}-OH$$

(b) $CH_3(CH_2)_6\overset{\overset{\displaystyle O}{\|}}{C}OH$

(d) 3-isopropylhexanoic acid

14. (a)

$$CH_3CH_2CH_2\overset{\overset{\displaystyle O}{\|}}{C}OH + HOCH_3 \longrightarrow CH_3CH_2CH_2\overset{\overset{\displaystyle O}{\|}}{C}OCH_3 + H_2O$$

(b)

(c)

$$H\overset{\overset{\displaystyle O}{\|}}{C}OH + HOCH_2CH_2CH_3 \longrightarrow H\overset{\overset{\displaystyle O}{\|}}{C}OCH_2CH_2CH_3 + H_2O$$

(d)

$$\underset{\text{CH}_3\text{COH}}{\overset{\displaystyle \underset{\|}{\text{O}}}{}} \quad + \quad \underset{\text{CH}_3\text{CHCH}_2\text{CH}_3}{\overset{\text{OH}}{\overset{|}{}}} \quad \longrightarrow \quad \underset{\text{CH}_3\text{COCHCH}_2\text{CH}_3}{\overset{\displaystyle \underset{\|}{\text{O}} \; \overset{\text{CH}_3}{\overset{|}{}}}{}} + \text{H}_2\text{O}$$

(e)

(f)

15. (a)

CH$_3$CH$_2$OH + HOC—⟨◯⟩

ethanol   benzoic acid

(b)

$$\underset{\underset{\text{CH}_3}{\overset{|}{}}}{\text{CH}_3\text{CHOCCH}_2\text{CH}_2\text{CH}_3} + \text{H}_2\text{O} \longrightarrow \underset{\underset{\text{CH}_3}{\overset{|}{}}}{\text{CH}_3\text{CHOH}} + \underset{}{\overset{\displaystyle \underset{\|}{\text{O}}}{}}\text{HOCCH}_2\text{CH}_2\text{CH}_3$$

isopropyl   butanoic
alcohol   acid

(c)

$$\text{CH}_3(\text{CH}_2)_6\text{CH}_2\text{OCCH}_3 + \text{H}_2\text{O} \longrightarrow \text{CH}_3(\text{CH}_2)_6\text{CH}_2\text{OH} + \text{HOCCH}_3$$

octanol   acetic
               acid

(d)

$$\underset{\underset{\text{CH}_3}{\overset{|}{}}}{\overset{\overset{\text{CH}_3}{\overset{|}{}}}{\text{CH}_3-\text{C}-\text{OCCH}_2\text{CH}_3}} + \text{H}_2\text{O} \longrightarrow \underset{\underset{\text{CH}_3}{\overset{|}{}}}{\overset{\overset{\text{CH}_3}{\overset{|}{}}}{\text{CH}_3-\text{C}-\text{OH}}} + \text{HOCCH}_2\text{CH}_3$$

t-butyl   propanoic
alcohol   acid

16. (a)

$$\text{CH}_3\text{OCCH}_3 + \text{NaOH} \longrightarrow \text{CH}_3\text{OH} + \text{NaOCCH}_3$$

278

(b) $CH_3CH_2OCCH_2CH_3$ + NaOH $\longrightarrow$ CH_3CH_2OH + $NaOCCH_2CH_3$

(c) 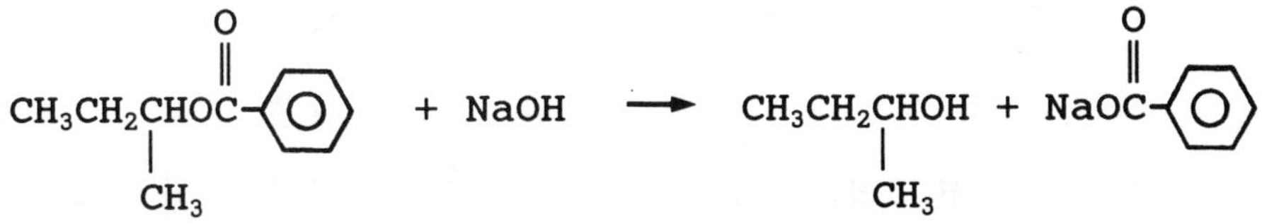 CH_3CH_2CHOC-⟨O⟩ + NaOH $\longrightarrow$ CH_3CH_2CHOH + $NaOC$-⟨O⟩
         $|$ $|$
         CH_3 CH_3

(d) CH_3CH_2OC-$COCH_2CH_3$ + 2NaOH $\longrightarrow$ $2CH_3CH_2OH$ + $NaOC$-$CONa$

Chapter 16

1. (a) primary
    (b) tertiary
    (c) primary
    (d) secondary
    (e) tertiary
    (f) primary

2. (a) tert-butylamine
    (b) N-ethyl-N-methylaniline
    (c) propylamine
    (d) isopropylmethylamine
    (e) dimethylethylamine
    (f) isopentylamine

3. e,a,b,c,d

4. (a)
    $$\begin{array}{c} CH_3 \\ | \\ CH_3CHCH_2NH_2 \end{array}$$ + HCl $\longrightarrow$ $$\begin{array}{c} CH_3 \\ | \\ CH_3CHCH_2NH_3Cl \end{array}$$
    isobutylammonium chloride

    (b) $(CH_3CH_2)_3N$ + HCl $\longrightarrow$ $(CH_3CH_2)_3NHCl$
    triethylammonium chloride

    (c) $(CH_3CH_2)_3N$ + CH_3CH_2Br $\longrightarrow$ $(CH_3CH_2)_4NBr$
    tetraethylammonium bromide

5. (a) pentamide
    (b) N,N-dimethylacetamide
    (c) N-phenylpropanamide
    (d) N-methyl-N-ethylhexanamide

6. (a) 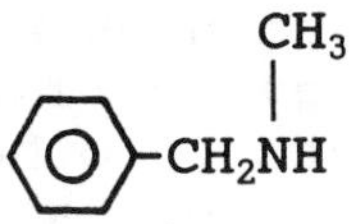 ⟨O⟩-CH_2NH
          $\quad\quad$ $|$
          $\quad\quad$ CH_3

    (b) $CH_3CH_2CH_2CH_2CNH_2$

    (d) $CH_3CH_2CH_2CH_2CH_2C$-NCH_2CH_3 with O and CH_2CH_3

    (e) CH_3CHCH_2OH with $NHCH_3$

279

(c) 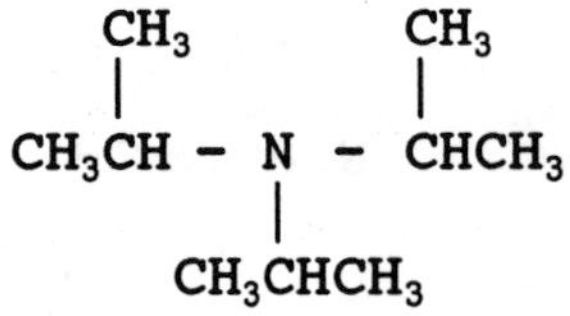
$$\underset{\underset{\displaystyle CH_3CHCH_3}{|}}{\overset{\overset{\displaystyle CH_3 \qquad\quad CH_3}{|\qquad\qquad|}}{CH_3CH - N - CHCH_3}}$$

(f) 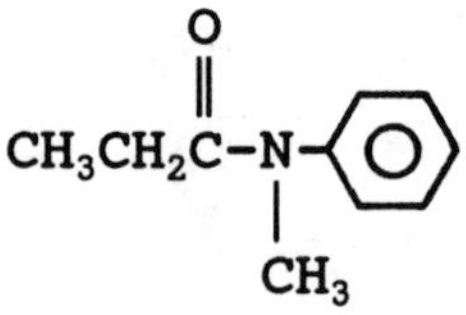
$$CH_3CH_2\overset{\overset{\displaystyle O}{\|}}{C}-\underset{\underset{\displaystyle CH_3}{|}}{N}\text{—}\langle O\rangle$$

7. (a) 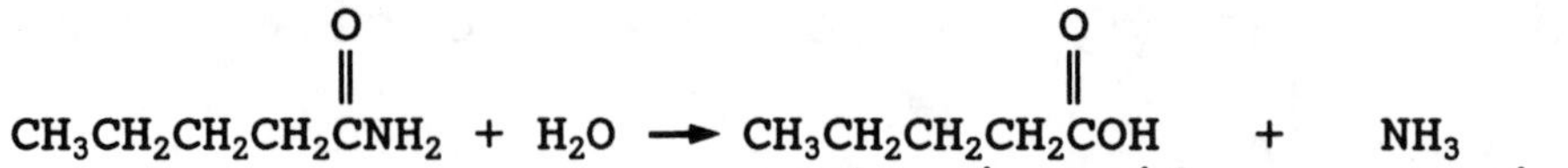 $CH_3CH_2CH_2CH_2\overset{\overset{\displaystyle O}{\|}}{C}NH_2 + H_2O \longrightarrow CH_3CH_2CH_2CH_2\overset{\overset{\displaystyle O}{\|}}{C}OH + NH_3$
    pentanoic acid   ammonia

   (b) 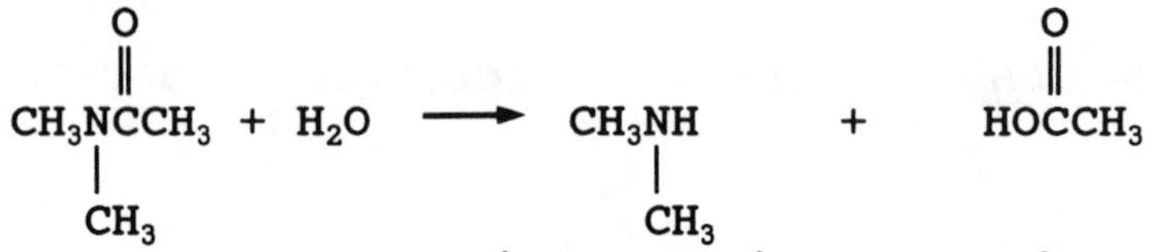 $CH_3\underset{\underset{\displaystyle CH_3}{|}}{N}\overset{\overset{\displaystyle O}{\|}}{C}CH_3 + H_2O \longrightarrow CH_3\underset{\underset{\displaystyle CH_3}{|}}{N}H + HO\overset{\overset{\displaystyle O}{\|}}{C}CH_3$
   dimethylamine   acetic acid

   (c)  $\langle O\rangle\text{—}NH\overset{\overset{\displaystyle O}{\|}}{C}CH_2CH_3 + H_2O \longrightarrow \langle O\rangle\text{—}NH_2 + HO\overset{\overset{\displaystyle O}{\|}}{C}CH_2CH_3$
   aniline   propanoic acid

   (d) 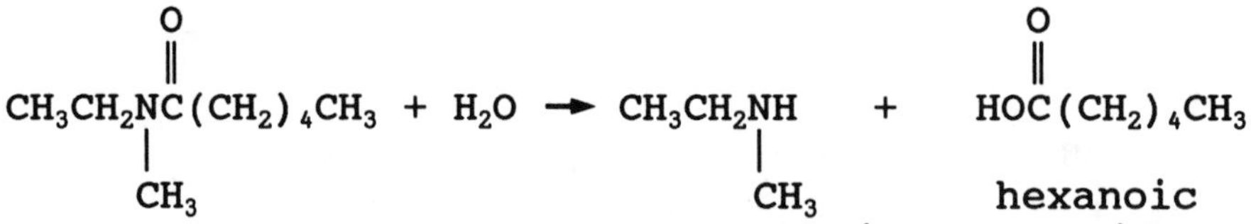 $CH_3CH_2\underset{\underset{\displaystyle CH_3}{|}}{N}\overset{\overset{\displaystyle O}{\|}}{C}(CH_2)_4CH_3 + H_2O \longrightarrow CH_3CH_2\underset{\underset{\displaystyle CH_3}{|}}{N}H + HOC(CH_2)_4CH_3$
   ethylmethylamine   hexanoic acid

8. Any two of the ring structures shown in Table 16.4 of the text are correct.

9. (a) epinephrine, norepinephrine   (f) nicotine
    (b) sodium pentothal   (g) heroin
    (c) caffeine   (h) morphine
    (d) barbiturate   (i) epinephrine
    (e) codeine

Chapter 17

1. Carbohydrates function as supporting structures of plants, energy storage molecules, immediate energy supplies for energy-requiring reactions, and components of other important biological molecules such as nucleic acids, glycolipids, and glycoproteins.

2. Monosaccharides are carbohydrates that cannot be broken into smaller units upon hydrolysis. Disaccharides will produce

two monosaccharides upon hydrolysis and polysaccharides will produce three or more monosaccharides upon hydrolysis.

3. (a) aldose, pentose   (c) ketose, hexose
    (b) ketose, heptose   (d) aldose, triose

4. (a)

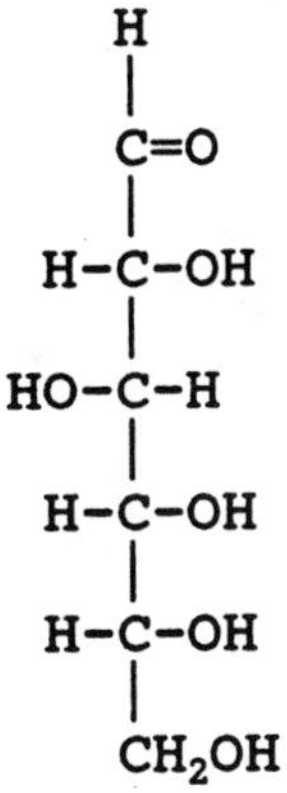

```
        H
        |
        C=O
        |
      H-C-OH
        |
      HO-C-H
        |
      H-C-OH
        |
      H-C-OH
        |
       CH₂OH
```

(b) α-glucose

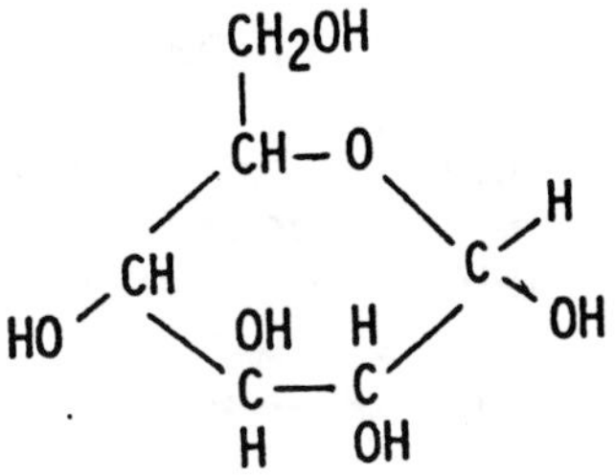

(c) β-glucose

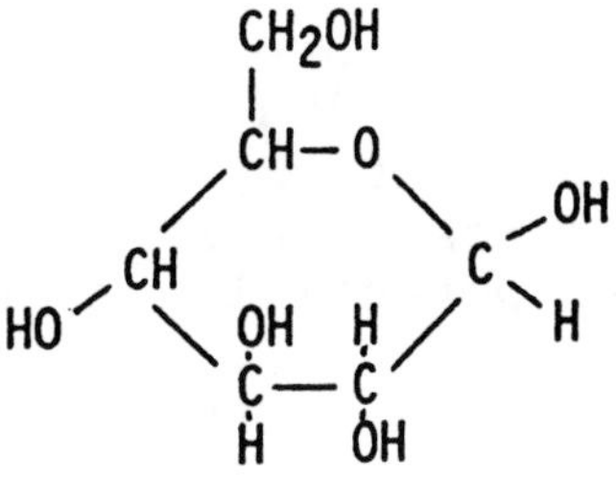

5. (a)

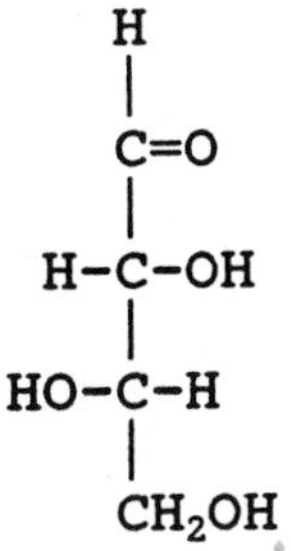

```
        H
        |
        C=O
        |
      H-C-OH
        |
      HO-C-H
        |
       CH₂OH
```

(b)

```
        H
        |
        C=O
        |
      H-C-OH
        |
      HO-C-H
        |
      HO-C-H
        |
       CH₂OH
```

(c)

```
        H
        |
        C=O
        |
      HO-C-H
        |
      HO-C-H
        |
      H-C-OH
        |
      HO-C-H
        |
       CH₂OH
```

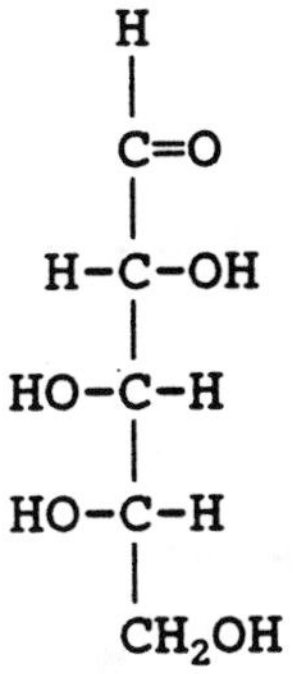

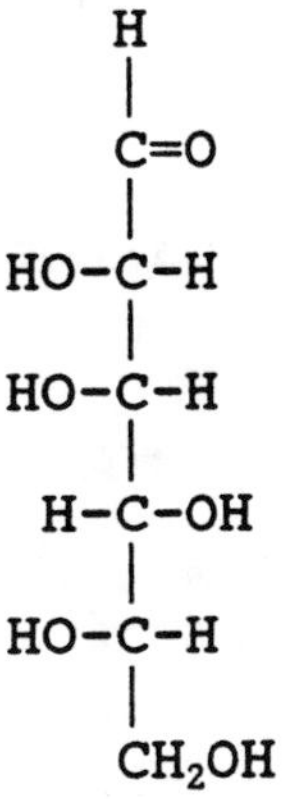

6. 32 aldoheptoses, 16 in the D-family and 16 in the L-family.

7.

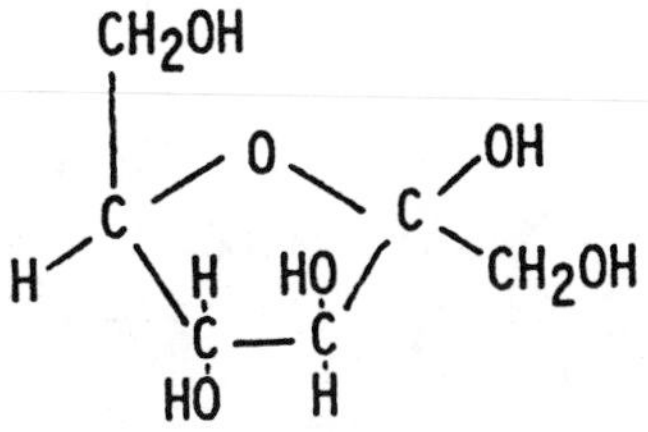

8. (a) pentose

   (b) alpha

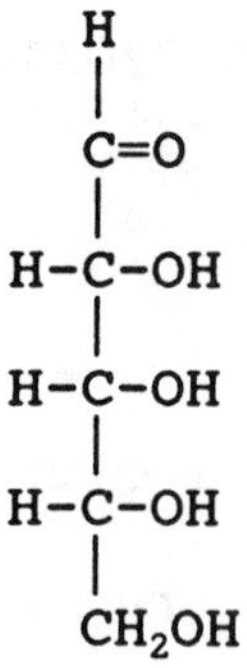

   (d) ribose

(c)
```
        H
        |
        C=O
        |
    H-C-OH
        |
    H-C-OH
        |
    H-C-OH
        |
      CH₂OH
```

9. Fructose, ribose, lactose, and maltose will all give a positive Benedict's test because they all have a free, or potentially free, aldehyde or ketone group.

10. (a)   sucrose   +   water   ⟶   glucose   +   fructose

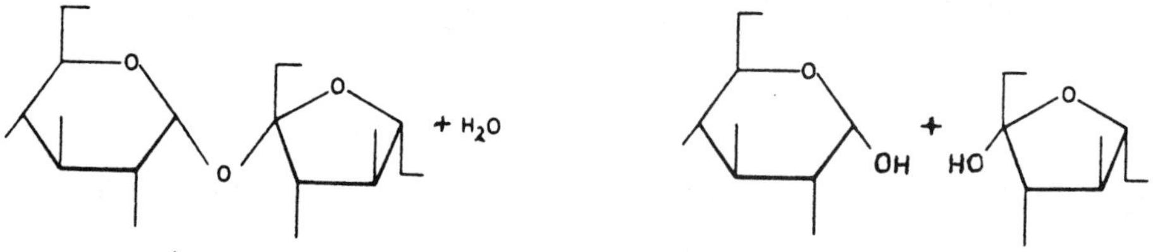

   (b)   maltose   +   water   ⟶   glucose   +   glucose

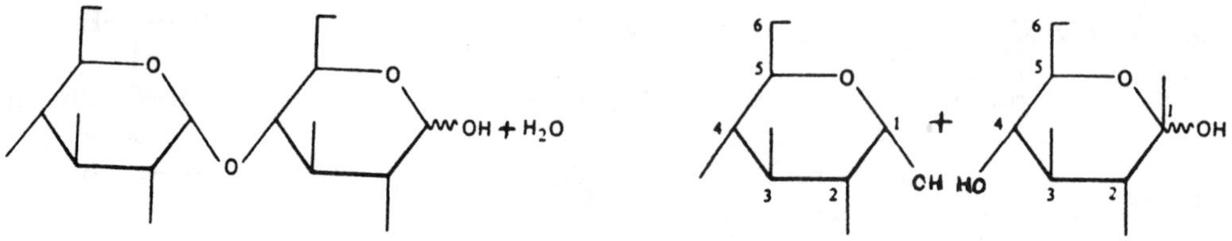

   (c)   lactose   +   water   ⟶   galactose   +   glucose

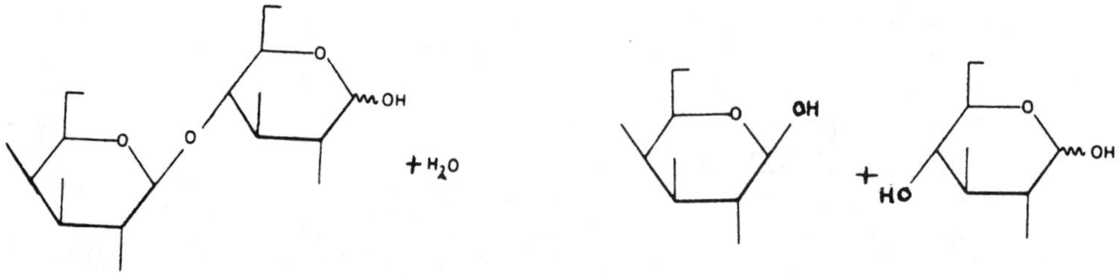

11. (a)   Amylose and amylopectin
   (b)   Amylose is a linear polymer of glucose, and amylopectin is a highly branched polymer of glucose.

(c) Place 15 drops of iodine test solution on a glass plate that is resting on a white sheet of paper. Add a small amount of starch solution to a flask that is in a water bath. Add several drops of hydrochloric acid solution to the starch to promote hydrolysis, and immediately remove a drop off this mixture and place it on the first drop of the iodine test solution. Take drops from the reaction mixture at fixed intervals and add them to the iodine test solution to monitor the progress of the reaction.

12. Glycogen is the storage form of glucose in animals. Its structure is very similar to the amylopectin portion of the starch molecule.

13. (a) carbon dioxide, water, and chlorophyll
    (b) chloroplasts of plant cells
    (c) $6CO_2 + 6H_2O + energy \longrightarrow C_6H_{12}O_6 + 6O_2$

Chapter 18

1. (a) A saponifiable lipid is one that can be hydrolyzed by a base, and a nonsaponifiable lipid is one that cannot be hydrolyzed by a base.
   (b) Simple triacylglycerols contain the same fatty acid on all three positions on the glycerol molecule, while mixed triacylglycerols contain two or more different fatty acids.
   (c) A simple lipid yields glycerol and three fatty acids upon hydrolysis. A compound lipid yields an alcohol, fatty acids, and other compounds upon hydrolysis.
   (d) Fat and oil are triacylglycerols, but a fat is a solid and an oil a liquid at room temperature.
   (e) A saturated fatty acid contains all carbon-to-carbon single bonds; an unsaturated fatty acid contains a carbon-to-carbon double bond; and a polyunsaturated fatty acid contains more than one carbon-to-carbon double bond.
   (f) An essential fatty acid is not produced by the body in sufficient amounts to maintain good health and must be provided in the diet. A nonessential fatty acid can be synthesized by the body.

2. The triacylglycerols in animal fats contain more saturated fatty acids than unsaturated fatty acids, whereas the triacylglycerols in vegetable oils contain more unsaturated fatty acids.

3. (a) saturated: myristic, lauric
       unsaturated: oleic, linolenic
   (b) oleic

4. Linolenic acid is one of the essential fatty acids that cannot be produced by the body but that are essential for good health.

5. $$CH_3(CH_2)_{14}\overset{\displaystyle O}{\overset{\displaystyle \|}{C}}OCH_2(CH_2)_{14}CH_3$$

6. The iodine number of a lipid gives information on how many unsaturated bonds are in the lipid's fatty acids.

7. (a) Hydrolysis is the addition of water to a molecule, forming two molecules; hydrogenation is the addition of hydrogen to the unsaturated bonds in a molecule.
    (b) Hydrolysis is the splitting of a lipid molecule by water, and saponification is the splitting of a lipid molecule by a base.

8. The unsaturated fatty acids in corn oil are partially hydrogenated. This produces a product with a higher melting point which is a solid at room temperature.

9. (a) The fats in butter undergo hydrolysis, producing the fatty acids that give butter a rancid taste and smell.
    (b) Refrigeration retards the rate of hydrolysis.
    (c) Butyric acid

10. (a) The nonpolar tails of the soap molecules dissolve in the grease, while the polar heads remain dissolved in the water. this breaks the grease into colloidal droplets that can be washed away by the water.
    (b) The soap molecules can disrupt and dissolve the bacterial cell membrane, thus killing the bacteria.

11. A detergent's solubility in water is not affected by metal ions or by changes in the pH of water, but the solubility of soap is affected by these factors.

12. (a) A triacylglycerol is a simple lipid containing glycerol and three fatty acids. Phospholipids are compound lipids containing an alcohol, fatty acids, a phosphate group, and a nitrogen-containing compound.
    (b) Phosphatidylcholine is a phosphatide containing choline, while sphingomyelin is a sphingolipid containing choline.
    (c) The difference between glycolipids and phospholipids is that glycolipids contain a sugar group in place of the phosphate group in the phospholipid molecule.

13. Phospholipids molecules contain a polar region and a nonpolar region and are good emulsifying agents. A fat molecule is nonpolar and is not a good emulsifying agent.

In fact, phospholipids are used to emulsify fats and make them water soluble

14. (a) Phosphoglycerides form cell membranes.

```
G - fatty acid
L
Y
C - fatty acid
E
R
O - phosphate - nitrogen compound
L
```

(b) Phosphatidylcholine is important in the transport and metabolism of fats, and is a source of inorganic phosphate for tissue formation.

```
G - fatty acid
L
Y
C - fatty acid
E
R
O - phosphate - choline
L
```

(c) Phosphatidylethanolamine is important in the clotting of blood and is a source of inorganic phosphate for tissue formation.

```
G - fatty acid
L
Y
C - fatty acid
E
R
O - phosphate - ethanolamine
L
```

(d) Sphingolipids are found in the brain and nervous tissue and form part of the myelin sheath.

```
S
P
H
I
N
G     -  fatty acid
O
S
I
N     -  phosphate  -  nitrogen compound
E
```

(e) Sphingomyelins form part of the myelin sheath around nerves.

```
S
P
H
I
N
G     -  fatty acid
O
S
I
N     -  phosphate  -  choline
E
```

(f) Glycolipids are components of brain and nerve tissue.

```
S                              G  -  fatty acid
P                              L
H                              Y
I                              C  -  fatty acid
N                              E
G     -  fatty acid            R
O                              O  -  sugar
S                              L
I
N     -  sugar
E
```

15. (a) steroids
    (b) cholesterol
    (c) They differ in one functional group: testosterone has an alcohol group where progesterone has a ketone group.

16. Cholesterol is a steroid component of cell membranes.

17. (a) Cholesterol is one of the materials that forms plaques in the arteries that result in atherosclerosis.
    (b) Cholesterol is essential to the normal functioning of the body and is the starting material from which the body synthesizes bile salts, sex hormones, and vitamin D.

18. (a) Cellular membranes provide a barrier between the aqueous solution of the cellular cytoplasm and the extracellular fluid. They ensure that all the

components of the cell are kept close together.
  (b) A nonpolar compound would most easily diffuse through the nonpolar cellular membrane.

Chapter 19

1. (a) Amino acids are carboxylic acids that have an amino group on the α-carbon. Proteins are polymers of amino acids.
    (b) A simple protein yields only amino acids upon hydrolysis, while a conjugated protein yields amino acids and other organic or inorganic substances upon hydrolysis.
    (c) A globular protein is soluble in water, fragile, and has a catalyzing or transporting function, while a fibrous protein is insoluble in water, physically tough, and has a structural or protective function.
    (d) A glycoprotein is a conjugated protein in which the prosthetic group is a carbohydrate. A lipoprotein is a conjugated protein in which the prosthetic group is a lipid.
    (e) Animal proteins are adequate proteins containing all essential amino acids while most vegetable proteins are inadequate proteins.

2. (a) glycine   (b) alanine   (c) leucine   (d) valine

3. (a)

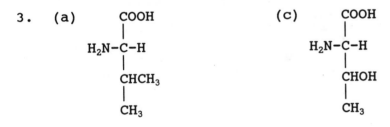

    (b)

4. (a) Adequate protein is protein that contains all the essential amino acids.
    (b) Most vegetable protein is inadequate protein. A vegetarian diet must contain a mixture of vegetable proteins to assure that all the essential amino acids are found in the diet.

5. Amino acids exist as dipolar ions and, therefore, have a strong attraction for one another which results in high melting points.

6. Blood proteins are amphoteric; they can act both as acids and bases in neutralizing additions of acid or base to the blood, and buffering the blood against changes in pH.

7. (a) Urease would migrate toward the positive pole.
   (b) Myoglobin would not migrate.
   (c) Chymotrypsin would migrate toward the negative pole.

8. At their isoelectric points protein molecules will be electrically neutral and tend to cluster together and precipitate out of solution.

9. (a) The primary structure of a protein is the sequence of amino acids in the molecule, and the secondary structure is the geometric arrangement formed by that sequence of amino acids.
   (b) A dipeptide contains two amino acids, and a polypeptide contains three or more amino acids connected by peptide bonds.
   (c) An alpha helix is a secondary configuration of a protein in which the amino acids form loops held together by hydrogen bonds. A beta configuration is a secondary structure of a protein in which several polypeptide chains are held together by hydrogen bonding in a zig-zag fashion.
   (d) Tertiary structure is the three-dimensional structure of a globular protein. Quaternary structure refers to the way in which polypeptide chains and prosthetic groups fit together in proteins containing more than one polypeptide chain.

10. (a) hydrogen bonding
    (b) hydrogen bonding, disulfide bridges, salt bridges, or hydrophobic interactions
    (c) hydrogen bonding
    (d) hydrogen bonding

11. The N-terminal end of a protein molecule is the end of the polypeptide chain with the free amino group. The C-terminal end of a protein molecule is the end of the polypeptide chain with the free carboxylic acid group.

12.

$$H_3N^+-CH-C(=O)-NH-CH-C(=O)-O^-$$

Left structure: H$_3$N$^+$–CH–C(=O)–NH–CH–C(=O)–O$^-$, with first CH bearing CH$_2$ connected to a benzene ring with OH (para), and second CH bearing CH$_2$–COOH.

Right structure: H$_3$N$^+$–CH–C(=O)–NH–CH–C(=O)–O$^-$, with first CH bearing CH$_2$–COOH, and second CH bearing CH$_2$ connected to a benzene ring with OH (para).

13. Phe-Ser-Ile   Ser-Phe-Ile
Phe-Ile-Ser   Ile-Phe-Ser
Ser-Ile-Phe   Ile-Ser-Phe

14. There is a greater number of disulfide bridges between the α-helixes of the keratin in hooves than in the keratins of wool.

15. Silk would be more disrupted by heat than would keratin because the polypeptide chains in silk are held together only by hydrogen bonds, whereas keratin's structure is held together by both hydrogen bonds and disulfide bridges.

16. (a) The native shape of a protein is the shape that is energetically most stable for that protein.
     (b) Hydrogen bonding, disulfide bridges, hydrophobic interactions, and salt bridges.

17. The R-group of cysteine contains a disulfide group that is easily oxidized to form a disulfide bond that stabilizes the native state.

18. (a) The interactions that hold the protein in its native shape are broken causing the protein to unwind.
     (b) In some cases, the process can be reversed.

19. The heat of pasteurization will coagulate the protein of the microorganisms, killing them.

Chapter 20

1. (a) Metabolism is the term given to all the enzyme catalyzed reactions in the body.
    (b) Catabolic reactions are reactions which produce cellular energy by breaking down molecules.
    (c) Anabolic reactions are biosynthetic reactions.
    (d) An enzyme is a biological catalyst.
    (e) A prosthetic groups is a cofactor that is tightly bound to the protein part of the enzyme.

2. (a) lipids
    (b) sucrose
    (c) cellulose
    (d) peptides
    (e) lactose
    (f) compounds with glycosidic
        linkages
    (g) proteins
    (h) esters

3. (a) oxidation-reduction reactions
    (b) transfer of amino groups
    (c) hydrolysis reactions
    (d) interconversion of isomers
    (e) reduction reaction
    (f) oxidation reaction
    (g) transfer of methyl groups
    (h) removal of hydrogen

4. (a) Step 1: The glucose-1-phosphate attaches to the
                 active site of the phosphoglucomutase
                 molecule.
        Step 2: The glucose-1-phosphate becomes activated.
        Step 3: Glucose-6-phosphate forms on the surface of
                 the phosphoglucomutase molecule.
        Step 4: The glucose-6-phosphate is released from the
                 active site on the phosphoglucomutase
                 molecule.
    (b) 6×10^4 molecules/hour: 1×10^{-19} moles/hour

5. (a) The lock-and-key analogy explains the specificity of
        enzyme action by stating that the conformation of the
        active site is specific to the geometry of the
        substrate molecule, just as the configuration of a lock
        is specific for one key.
    (b) The induced-fit theory modifies the lock-and-key theory
        by stating that in some cases the active site of the
        enzyme does not exactly conform to the substrate, but
        the substrate itself induces the flexible enzyme
        molecule to take on a shape that conforms to the
        substrate molecule, just as a hand induces a fit when
        it slips into a glove.

6. Pepsin has an optimum activity at pH 1.5, and its activity
    will decrease as the pH increases. The small intestines
    have an alkaline pH and, as a result, the activity of pepsin
    is very low.

7. Boiling water will denature the protein in microorganisms,
    thereby killing them and sterilizing whatever is boiled in
    the water.

8. (a) vitamin B_{12} (c) vitamin D   (e) niacin
    (b) vitamin K   (d) vitamin C

9. (a) Vitamins A and D are fat soluble vitamins and are
         stored in body fat, but vitamins C and B_1 are water
         soluble vitamins and are readily excreted by the body.
    (b) Both a deficiency and an excess of vitamin D cause a
         disruption in calcium metabolism. A deficiency will
         cause rickets, and an excess can cause growth
         retardation. Because excess vitamin D is stored in the
         fat and not readily excreted, a toxic effect can
         readily occur.

10. nicotinamide adenine dinucleotide
    nicotinamide adenine dinucleotide phosphate
    flavin adenine dinucleotide.

11. coenzyme A

12. (a) Vitamin D is a fat soluble vitamin and, when eaten in
         excess of daily requirements, is stored in fatty
         tissue. In high concentrations it can be harmful to
         the health of a child. By eating so many foods that
         are fortified with vitamin D, a child could easily
         exceed his daily requirements.
    (b) Fortification of foods with vitamins should be
         carefully controlled to protect the consumer from an
         excess of fat-soluble vitamins, which is potentially as
         harmful as a deficiency.

13. (a) An allosteric enzyme is an enzyme that controls the
         reaction rate of a multienzyme system.
    (b) If all possible reactions occurred in a cell at a high
         rate, the cell would certainly die. Allosteric enzymes
         allow the cell to control the rate of reactions,
         producing only those products required by the cell.
    (c) The regulatory molecule can either inhibit or increase
         the activity of the allosteric enzyme by attaching to
         the allosteric site on the enzyme molecule.

14. Hormones regulate cellular processes by attaching to the
    cell membrane triggering changes within the cell or they
    enter the cell and migrate to the nucleus where they
    activate specific genes.

15. (a) Prostaglandins raise and lower blood pressure, regulate
         gastric secretions, cause uterine contractions, and
         dilate the opening of the air passages to the lungs.
         They also produce fever and inflammation.
    (b) Prostaglandins are fatty acids produced in most cells
         and, like cyclic AMP, they carry out the messages that
         the cells receive from hormones.

16. The molecule that the liver synthesizes from Parathion
    inhibits the enzyme cholinesterase in the nerves.

Cholinesterase catalyzes the hydrolysis of acetylcholine. When cholinesterase is inhibited, the resulting overstimulation of the nerve cells by acetylcholine cause irregular heart rhythms, convulsions, and death.

17. An antimetabolite is a chemical substance that inhibits enzyme activity. Sulfanilamide and penicillin are antimetabolites.

18. A noncompetitive inhibitor combines reversibly with a portion of the enzyme molecule that is essential to enzyme function. It can affects all enzyme molecules. A competitive inhibitor competes with the substrate for the active site on the molecule. Sometimes it is effective in attaching to the active site and sometime it is not, so an competitive inhibitor will affect only a portion of the enzyme molecules.

19. Lead ions bind irreversibly to the sulfhydryl groups on enzyme molecules, inactivating them. This disrupts normal metabolism and causes the symptoms of lead poisoning.

Chapter 21

1. (a)

(b) A high-energy phosphate bond is one that yields twice as much energy upon hydrolysis as an ordinary phosphate bond.

(c) ATP molecules supply the energy required for the anabolic reactions occurring within the cell.

2. Digestion of carbohydrates begins in the mouth as the teeth break up the food and an enzyme in the saliva begins the breakdown of the starch to maltose. In the intestines enzymes completely hydrolyze the polysaccharides and disaccharides to glucose, fructose, and galactose. These monosaccharides are absorbed through the intestinal walls into the bloodstream and are carried to the liver to be converted to glucose. Digestion of lipids begins in the small intestine where bile salts help emulsify fats and allow enzymes to hydrolyze the fats into glycerol, fatty acids, and mono- and diacylglycerols. These products of hydrolysis cross the intestinal wall and enter the lymph system. Protein digestion starts in the stomach and continues in the small intestine where the proteins are hydrolyzed into amino acids and small peptides. These products are absorbed through the intestinal wall and enter the bloodstream.

3. The level of glucose in the blood is carefully controlled by the liver and by several hormones that can cause the liver to take up glucose or to produce it. After a meal, glucose enters the blood in large amounts. Insulin is produced by the pancreas to promote the uptake of glucose by the liver and muscle cells, where it is converted into glycogen. As the glucose in the blood is utilized by the tissues, the level of glucose in the blood drops. Glucagon, another hormone produced by the pancreas, then causes an increase in the rate of glycogenolysis in the liver and, therefore, an increase in the concentration of glucose in the blood. In this way the concentration of glucose in the blood is maintained at a fairly constant level throughout the day.

4. A diet for a person suffering from hypoglycemia should be high in protein and should contain a moderate amount of fat and carbohydrate. No meal or snack should be high in carbohydrate to avoid an overstimulation of the pancreas.

5. (a) glycolysis
    (b) glucose + 2ADP + P_i $\longrightarrow$ 2 lactic acid + 2ATP + 2H_2O
    (c) The anaerobic stage occurs in the cytoplasm, and the aerobic stage in the mitochondria.
    (d) aerobic stage
    (e) The two series of reactions are the citric acid cycle and the electron transport chain. An acetyl CoA enters the citric acid cycle and is oxidized to two molecules of carbon dioxide and four pairs of hydrogens that are bound to hydrogen carriers. These hydrogens then enter the electron transport chain, which produces water and

energy in the form of ATP.

(f) $acetyl\ CoA + GDP + 2H_2O \longrightarrow 2CO_2 + GTP + 11ATP$
$+ coenzyme\ A$

6. (a) $glucose + ATP \longrightarrow glucose\text{-}6\text{-}phosphate + ADP$
$fructose\text{-}6\text{-}phosphate + ATP \longrightarrow fructose\text{-}1,6\text{-}phosphate$

(b) $1,3\text{-}diphosphoglycerate + ADP \longrightarrow 3\text{-}phosphoglycerate$
$+ ATP$
$phosphoenolpyruvate + ADP \longrightarrow pyruvate + ATP$

(c) $glucose\text{-}6\text{-}phosphate \longrightarrow fructose\text{-}6\text{-}phosphate$
$glyceraldehyde\text{-}3\text{-}phosphate \longrightarrow dihydroxyacetone\ phosphate$
$3\text{-}phosphoglycerate \longrightarrow 2\text{-}phosphoglycerate$

7. (a) $\dfrac{300\ kcal}{1\ hr} \times \dfrac{1\ mol}{7.3\ kcal} = \dfrac{41\ mol\ ATP}{1\ hr}$

(b) $glucose \longrightarrow 2\ pyruvate + 2ATP$
$2\ pyruvate \longrightarrow 2\ acetyl\ CoA + 2NADH + H^+(6\ ATP)$
$2\ acetyl\ CoA \longrightarrow 4CO_2 + 24ATP$

oxidation of 1 mol of glucose produces 32 mol ATP

$\dfrac{41\ mol\ ATP}{1\ hr} \times \dfrac{1\ mol\ glucose}{32\ mol\ ATP} \times \dfrac{180\ g}{1\ mol\ glucose} = \dfrac{230\ g}{1\ hr}$

8. (a) When your muscles first begin strenuous exercise, the energy required for contraction is supplied by aerobic respiration. But as contraction continues, the body cannot supply oxygen fast enough to the muscles tissues and, as a result, the energy for continued contraction is supplied by glycolysis. Lactic acid begins to build up in the muscle tissue until a concentration of lactic acid is reached that impairs muscle function, causing muscle fatigue.

(b) When the exertion is over, you continue to gasp for breath to supply the oxygen that is needed to oxidize the lactic acid that was produced during the exercise.

9. In yeast cells glycolysis results in the formation of ethanol and carbon dioxide while in animal cells, lactic acid and water are formed.

10. In order for fats to be oxidized, they must first be hydrolyzed to glycerol and fatty acids. The glycerol enters the glycolysis pathway, and the fatty acids are oxidized by the process of beta oxidation to acetyl CoA, which then enters the citric acid cycle.

11. (a) The fatty acid is joined with coenzyme A.
(b) The acetyl CoA can enter the citric acid cycle or can

be used in the synthesis of compounds used by the cell.
- (c) This process creates an alcohol on the beta carbon which is then oxidized to a ketone.

12. (a)

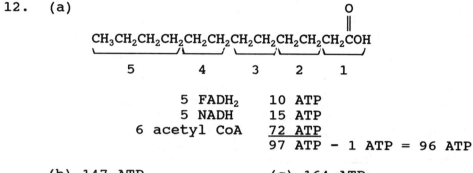

$$CH_3CH_2CH_2CH_2CH_2CH_2CH_2CH_2CH_2CH_2CH_2\overset{\overset{\displaystyle O}{\|}}{C}OH$$

    5   4   3   2   1

        5 $FADH_2$ 10 ATP
        5 NADH   15 ATP
    6 acetyl CoA   <u>72 ATP</u>
                  97 ATP − 1 ATP = 96 ATP

   (b) 147 ATP   (c) 164 ATP

13. Acetyl CoA is important in the citric acid cycle; it can be used in the biosynthesis of compounds required by the cell; and if present in excess, it can be used to synthesize ketone bodies.

14. (a) Ketosis occurs when the level of ketone bodies in the blood exceeds the amount that can be used by the tissues and is, therefore, excreted in the urine.
- (b) Liver damage, diabetes mellitus, starvation, or diets can cause a decrease in glucose metabolism and an increase in fat metabolism with the accompanying increase in ketone bodies.
- (c) Ketosis causes acidosis.

15. Amino acids are used to synthesize tissue protein for formation or repair of cells and to synthesize other amino acids, enzymes, hormones, and antibodies. Amino acids are used to synthesize nonprotein nitrogen-containing compounds such as nucleic acids or heme groups. Amino acids can also be broken down for energy.

16.

    COOH   COOH   COOH   COOH
    | | | |
H−C−NH_2 +   C=O   ⟶   C=O + H−C−NH_2
    | | | |
  CH−CH_3 CH_3 CH−CH_3 CH_3
    | |
    OH   OH

17.

        COOH   COOH
        | enzymes |
    H−C−NH_2 ⟶   C=O + NH_3
        | H_2O |
       CH_3 CH_3

18. The ammonia produced by the oxidative deamination of proteins is converted by cells in the liver to urea through the reactions of the urea cycle. The urea enters the blood, is carried to the kidneys, and is excreted.

19. Coenzyme A

Chapter 22

1. (a) A nucleotide is composed of a nitrogen-containing base, a five-carbon sugar, and a phosphate group.
   (b) Free nucleotides in the cell participate in biosynthetic reactions, serve as coenzymes, and are important in the transport of energy from energy-releasing reactions to energy-requiring reactions.
   (c) A nucleic acid is a polymer of many nucleotides.

2. Thymine and cytosine are pyrimidines, and adenine and guanine are purines.

3. (a)

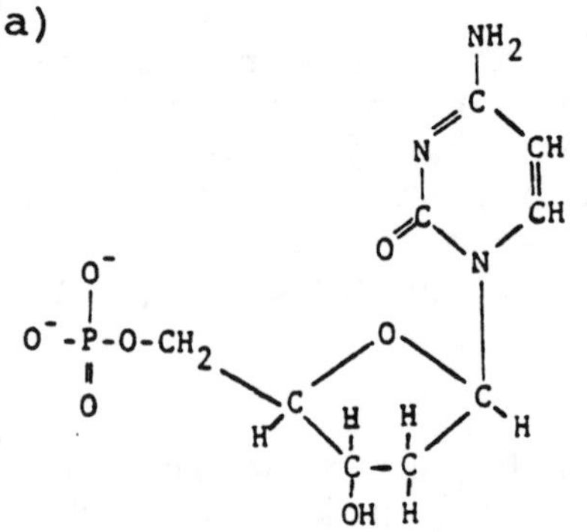

   (b)

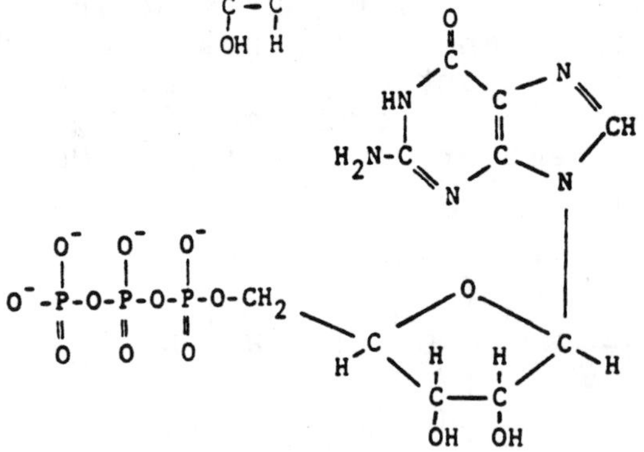

4. The two polynucleotide chains are held together by hydrogen bonding between the bases.

5. Thymine and guanine both have carbonyl groups on the first position. A carbonyl group cannot hydrogen bond with another carbonyl group. Also thymine and guanine have an

amino group in the middle of the three positions that form hydrogen bonds. An amino group cannot form hydrogen bonds with another amino group. So thymine and guanine cannot form hydrogen bonds with one another in either the first or second position that form hydrogen bonding in DNA.

6. The pairing of bases on the two polynucleotide chains of DNA is very specific: thymine pairs only with adenine, and cytosine with guanine. Therefore, the order of bases on each chain is not identical; where there is a thymine on one chain, there will always be an adenine on the other. That is, the chains are complementary.

7. For a DNA molecule to replicate, the two strands of the helix unwind, and each strand serves as a template for the synthesis of a new strand of DNA, forming two daughter DNA molecules each having an original DNA strand and one newly-made strand.

8. Three types of RNA are messenger RNA, transfer RNA, and ribosomal RNA. mRNA serves as the template for protein synthesis; tRNA transports the amino acids to the ribosomes for use in protein synthesis, and rRNA along with protein forms the ribosomes, the sites of protein synthesis.

9. (a) Genetic information is carried in the sequence of bases on the DNA molecule.
   (b) 20 amino acids
   (c) methionine –AUG and tryptophan - UGG

10. (a) TAC GGT GTA CAT AAC TTG GGG TAA GAC ACT
    (b) Met-Pro-His-Val-Asn-Pro-Pro-Ile-Leu

11. (a) DNA in eukaryotic cells contains exon segments that code for amino acids and intron segments that do not code for amino acids.
    (b) The DNA is transcribed to hnRNA which is spliced by special enzymes to take out all of the intron segments leaving the exon segments to form the mRNA.

12. (a) Protein synthesis occurs in the ribosomes.
    (b) 1) mRNA is synthesized in the nucleus and migrates to the cytoplasm.
    2) The smaller subunit of the ribosome combines with the mRNA, three initiation factors, and a tRNA carrying the start amino acid.
    3) This complex joins with the larger half of the ribosome. The growth of the amino acid chain takes place through a series of repeated steps in which the anticodon region of a charged tRNA is matched with the codon on the mRNA and peptide bonds are formed between the amino acids on the tRNAs.

4) The elongation of the chain stops when the terminal codon is reached on the mRNA.

13. The piece of DNA that would code for cortisone is synthesized or isolated from human DNA. It is inserted into a plasmid which is then inserted into E. coli. The E. coli are grown in large vats and produce the cortisone along with bacterial proteins. The cortisone is then isolated and purified for use.

14. (a) A mutation is any change in the base sequence on DNA that results in a change in the amino acids on a polypeptide chain.
    (b) Mutations can occur when one base is incorrectly inserted into the DNA molecule, when segments of the DNA are deleted, or additional segments of DNA are added to the gene.
    (c) The nucleus of a cell contains special enzymes that continually read the DNA and correct base errors and remove additional bases or connect broken pieces of DNA to prevent mutations.

15. (a) PKU is caused by an error in the gene that codes for phenylalanine hydroxylase.
    (b) PKU disrupts the metabolism of the amino acid phenylalanine and the metabolites that accumulate in the blood can over time causing permanent brain damage.
    (c) The effects of PKU can be minimized by feeding the infant special formula that doesn't contain phenylalanine.
    (d) Possibly, if the DNA that codes for phenylalanine hydroxylase could be inserted into human liver cells that then could be injected into the patient's liver.

NOTES

NOTES

NOTES

NOTES

NOTES

NOTES

NOTES

NOTES

NOTES

NOTES

NOTES

NOTES

NOTES